THE HISTORICAL ROOTS OF
ELEMENTARY MATHEMATICS

THE HISTORICAL ROOTS OF ELEMENTARY MATHEMATICS

Lucas N. H. Bunt
Late of Arizona State University

Phillip S. Jones
Professor Emeritus, University of Michigan

Jack D. Bedient
Arizona State University

DOVER PUBLICATIONS, INC., NEW YORK

Published in Canada by General Publishing Company, Ltd., 30 Lesmill Road, Don Mills, Toronto, Ontario.

This Dover edition, first published in 1988, is an un-abridged republication of the work originally published by Prentice-Hall, Inc., Englewood Cliffs, New Jersey, in 1976. Authors' corrections have been incorporated into this edition.

Manufactured in the United States of America
Dover Publications, Inc., 31 E. 2nd Street, Mineola, N.Y. 11501

Library of Congress Cataloging-in-Publication Data

Bunt, Lucas N. H. (Lucas Nicolaas Hendrik)
 The historical roots of elementary mathematics / Lucas N. H. Bunt, Phillip S. Jones, Jack D. Bedient.
  p.   cm.
 Reprint. Originally published: Englewood Cliffs, N.J. : Prentice-Hall, c1976. With authors' corrections.
 Includes bibliographical references and index.
 ISBN 0-486-25563-8 (pbk.)
 1. Mathematics—History. I. Jones, Phillip S. II. Bedient, Jack D., 1926–   . III. Title.
[QA21.B95 1988]
510'.93—dc19   87-25059
                                               CIP
                                                AC

To Lies
   Helen
   Mary

CONTENTS

PREFACE

This book has several unique characteristics which have developed from the experiences of the authors in teaching historical material.

The book is topical rather than comprehensive. Although the reader may not find in it some specific information that would appropriately appear in a reference book, the topics introduced are developed with sufficient depth so that the student can actually carry out the mathematical tasks in an authentically historical setting. He can do long division like the ancient Egyptians, solve quadratic equations like the Babylonians, and study geometry just as the student in Euclid's day. To get involved in the same processes and problems as the ancient mathematicians and to effect solutions in the face of the same difficulties they faced is the best way to gain appreciation of the intelligence and ingenuity of the scholars of early times. The authors have found that students enjoy this intensive approach to study of the history of mathematics and that they gain in their understanding of today's mathematics through analyzing older and alternative approaches.

The book covers the historical roots of elementary mathematics: arithmetic, algebra, geometry, and number systems. It omits many of the topics from later periods of mathematics. Many of the concepts under development during these later periods are beyond the reach of undergraduate mathematics students, and discussions of such concepts at a superficial level are almost meaningless. The student with a good background in high school mathematics can study from this book with profit; substantial segments (for example, Babylonian, Egyptian, Greek, and other numeration systems and computational algorithms) are within the range even of many junior high school students.

Since the origins of most of the topics included in the elementary and high school mathematics curriculum are discussed, the book is particularly appropriate for the prospective teacher of mathematics. The authors' experience has shown that this material can provide a basis for a three-semester-hour course in the history of mathematics for mathematics majors and for prospective secondary school teachers. The book is suitable also for courses for prospective elementary school teachers (particularly Chapters 1, 2, 3, 6, and 8), as enrichment material for high school students, and for the edification of the general reader. It is the sincere wish of the authors that the book will make the history of mathematics available to a broader audience than heretofore.

Much of the material in the book is derived from a Dutch text, *Van Ahmes tot Euclides* (Wolters, Groningen), written by Dr. Lucas N. H. Bunt, then at the University of Utrecht, The Netherlands, and the following collaborators: Dr. Catharina Faber-Gouwentak, Sister E. A. de Jong, D. Leujes, Dr. H. Mooij, and Dr. P. G. J. Vredenduin. The present book contains many modifications and extensions contributed by Phillip S. Jones, including the latter part of Chapter 6, the first part of Chapter 7, and most of Chapter 8. Much of the material in his booklet *Understanding Numbers: Their History and Use* has been incorporated. Jack D. Bedient joined in the final organization of the manuscript.

The authors are greatly indebted to Professor Bruce E. Meserve, whose many suggestions contributed immeasurably to the improvement of the manuscript. His help and encouragement have kept the project alive to completion. They also wish to express their gratitude to the staff of Prentice-Hall for their helpfulness. The services of Anna Church and Emily Fletcher in typing the manuscript are much appreciated.

Tempe, Arizona   Lucas N. H. Bunt
Ann Arbor, Michigan   Phillip S. Jones
Tempe, Arizona   Jack D. Bedient

THE GREEK ALPHABET

α	alpha	ν	nu
β	beta	ξ	xi
γ	gamma	o	omicron
δ	delta	π	pi
ϵ	epsilon	ρ	rho
ζ	zeta	σ	sigma
η	eta	τ	tau
θ	theta	υ	upsilon
ι	iota	φ	phi
κ	kappa	χ	chi
λ	lambda	ψ	psi
μ	mu	ω	omega

THE HISTORICAL ROOTS OF
ELEMENTARY MATHEMATICS

1

EGYPTIAN MATHEMATICS

1-1 PREHISTORIC MATHEMATICS

The earliest written mathematics in existence today is engraved on the stone head of the ceremonial mace of the Egyptian king Menes, the founder of the first Pharaonic dynasty. He lived in about 3000 B.C. The hieroglyphics on the mace record the result of some of Menes' conquests. The inscriptions record a plunder of 400,000 oxen, 1,422,000 goats, and 120,000 prisoners. These numbers appear in Figure 1-1, together with the pictures of the ox, the goat, and the prisoner with his hands behind his back. Whether Menes exaggerated his conquests is interesting historically but does not matter mathematically. The point is that even at this early date, man was recording very large numbers. This sug-

Figure 1-1 From a drawing (Plate 26B) in J. E. Quibell, *Hierakonopolis*, London, 1900.

gests that some mathematics was used in the centuries before 3000 B.C., that is, before the invention of writing (the prehistoric period).

There are two ways of learning about the mind and culture of prehistoric humans. We have learned about them through the discovery of ancient artifacts, which were found and interpreted by archaeologists. We have also learned about prehistoric civilization by observing primitive cultures in the modern world and by making inferences as to how prehistoric thought and customs developed. In our study of the development of ideas and understandings, both approaches are useful.

One of the most exciting archaeological discoveries was reported in 1937 by Karl Absolom as a result of excavations in central Czechoslovakia. Absolom found a prehistoric wolf bone dating back 30,000 years. Several views are shown in Figure 1-2. Fifty-five notches, in

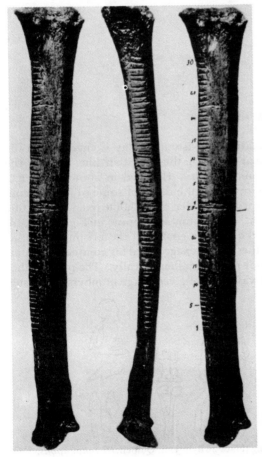

Figure 1-2 Prehistoric Wolf Bone. (From *London Illustrated News,* October 2, 1937.)

groups of five, are cut into the bone. The first 25 are separated from the remaining notches by one of double length. Although we do not know how this bone was notched, the most plausible explanation is that some prehistoric man deliberately cut it. Perhaps he was recording the number of a collection, possibly of skins, of relatives, or of days since an event. It is reasonable to assume that he made a notch for each object in the collection that he was counting. If this interpretation is correct, then we can recognize in this prehistoric record rudimentary versions of two important mathematical concepts. One is the idea of a *one-to-one correspondence* between the elements of two different sets of objects, in this case between the set of notches on the bone and the set of whatever the prehistoric man was counting. The other is the idea of a *base* for a system of numeration. The arrangement of the notches in groups of 5 and of 25 indicates a rudimentary understanding of a base 5 system of numeration.

Anthropological studies reinforce our belief in the existence of prehistoric number ideas. A study of the western tribes of the Torres Straits, reported by A. C. Haddon in 1889, describes a tribe that had no written language which counted as follows: 1, *urapun*; 2, *okosa*; 3, *okosa-urapun*; 4, *okosa-okosa*; 5, *okosa-okosa-urapun*; 6, *okosa-okosa-okosa*. Everything greater than 6 they called *ras*. A student of modern mathematics would recognize in this system of counting the beginnings of a base 2 numeration system. If a Torres Strait native had recognized this idea, however, he would have used a different word for 4 and would not have said *ras* for numbers greater than 6.

A. Seidenberg has recently published a theory of the origin of counting (see reference 12 at the end of the chapter). He believes that counting was invented for use in early religious rituals. Many studies of primitive tribes as well as early Babylonian religious writings are cited which indicate that participants in religious rituals were called into the ritual in a definite order and that counting developed in connection with specifying this order. In his studies, Seidenberg found *2-counting* to be the earliest counting that he could detect. This seems to indicate that the counting of the Torres Straits natives is consistent with a method of counting that was in use thousands of years earlier.

These two types of prehistoric number ideas, matching and counting, correspond to two different approaches to number which are common both in modern life and in modern education. One of these is the approach through the ideas of set and one-to-one correspondence between sets that have developed since the work of *Georg Cantor* (1845-1914) in the latter part of the nineteenth century. This treatment is sometimes referred to as a *cardinal* approach to number. At about the time that Cantor was developing the beginnings of modern set theory, *Giuseppe Peano* (1858-1932) was attempting to axiomatize the natural numbers and their arithmetic. To do this, he stated a set of five axioms. One of these axioms is that every natural number has a successor. Such a treatment is called an *ordinal* approach to number. It emphasizes the counting idea, in contrast to the matching idea stressed in the cardinal approach to number. These two approaches can be shown to be equivalent to each other, but our purpose here is merely to point out the antiquity of the underlying ideas which have recently been organized into important modern mathematical systems.

Evidences of other prehistoric mathematical ideas are not hard to find. One can read into primitive cave paintings some ideas of proportion and symmetry as skilled artists produced remarkably realistic drawings of animals and hunters. Ideas of number and one-to-one correspondence appear in connection with stickmen and four-legged animals. Elaborate geometric designs can be found on prehistoric pottery. Prehistoric drawings showing different views of a wagon and horses have been found in Europe. Sketches from the time of ancient Babylonia that might be plans of a building, perhaps a temple, have been unearthed. What appears to be a decimally divided ruler has been unearthed at Mohenjo-Daro in Pakistan. Interesting as these archeological findings are, from a mathematical point of view we shall find a study of the historic period to be more profitable. Let us therefore turn our attention to the earliest written mathematics, that of the Egyptians and the Babylonians.

1-2   THE EARLIEST WRITTEN MATHEMATICS

Although monuments, inscriptions, and Menes' mace record the earliest written numbers, most of our knowledge of Egyptian mathematics comes from writings on papyrus. Papyrus is a paperlike substance made from the papyrus plant, which grows along the Nile River. From these writings we learn that mathematics was studied in Egypt as early as 2000 years before Christ. Why in Egypt?

Herodotus (about 450 B.C.) observed that the Egyptians were forced to reset the boundary markers of their fields after the spring flooding of the Nile destroyed them. For that purpose surveyors were needed with some practical knowledge of simple arithmetic and geometry. Many of their computations remain. However, it is typical of Egyptian mathematics that arithmetic processes and geometric relations are described without mention of the underlying general principles. Thus, we know how the Egyptians performed computations, but we can only guess at how they developed their methods. We look for the reasoning behind their methods by deciphering and studying the detailed solutions of many examples.

The Greek mathematician *Democritus* (about 460-370 B.C.) appreciated the mathematical knowledge of the Egyptians as highly as his own achievements in this field. He writes: "In the construction of lines with proofs I am surpassed by nobody, not even the so-called rope stretchers of Egypt." By rope stretchers he probably meant surveyors, whose main instrument was the stretched rope. Figure 1-3 shows a statue of a rope stretcher with his coil of rope.

Figure 1-3 Surveyor holding coiled rope. Statue of Senemut, architect of Queen Hatshepsut, in the Louvre. (From *Histoire Générale des Sciences*, vol. 1, edited by René Taton, Presses Universitaires de France, 1957.)

The oldest known Egyptian mathematical texts contain mostly problems of a practical nature, such as computing the capacity of a granary, the number of bricks needed for the building of a store, or the stock of grain necessary for the preparation of a certain amount of bread or beer.

The *Rhind papyrus* is our best source of information about Egyptian arithmetic. It is named after an Englishman, A. Henry Rhind, who bought the text in Luxor in 1858 and sold it to the British Museum, where it is displayed. This papyrus, copied by a scribe, *Ahmes*, and sometimes called by his name, dates from about 1650 B.C., although, according to the writer, it had been taken from an older treatise written between 2000 and 1800 B.C. The text contains about 80 problems. Besides including solutions for many practical questions, some of which include geometric concepts, it contains a number of problems that are of no practical importance. We get the impression that the author posed himself problems and solved them for the fun of it.

There are four other, smaller Egyptian mathematical writings of some importance: the *Moscow papyrus*, the *Kahun papyrus*, the *Berlin papyrus*, and the *leather roll*. There are many small fragments and commercial papyri scattered around the world, but they furnish only slight information about Egyptian mathematics.

No definite place of discovery is known for the Moscow papyrus. It is named after the city where it is kept. A start was made on deciphering it in 1920. The complete document was published in 1930. The papyrus contains about 30 worked-out problems. Figure 1-9 (see page 38) contains a picture of a part of the papyrus.

About 1900 an Englishman discovered a papyrus in Kahun, hence its name. This papyrus contains applications of the arithmetic methods described in the Rhind papyrus, but it contains little more of importance.

Since in the course of years the leather roll had completely dried up and become hardened, it was extremely difficult to unfold it without destroying the text. Modern chemical processes have made it possible to soften and preserve it. The leather roll, which is displayed in the British Museum, will be discussed and shown in Section 1-9.

1-3   NUMERICAL NOTATION

Egyptian numerical notation was very simple. It used symbols for 1, 10, 100, ..., 1,000,000. In *hieroglyphics* these symbols were:

$$1 = | \qquad\qquad 1,000 = ⚡$$
$$10 = ∩ \qquad\qquad 10,000 = 〗$$
$$100 = ℮ \qquad\qquad 100,000 = ⌒$$
$$1,000,000 = 𐦂.$$

The symbol for 1000 was a lotus flower, for 10^4 a finger with a bent tip, for 10^5 a tadpole, and for 10^6 a man with his arms uplifted. Look back at Menes' mace in Figure 1-1 for examples of these symbols.

The numbers 2 through 9 were represented by two, three, ... , nine vertical dashes, as follows:

$$2 = || , \quad 3 = ||| , \quad \ldots, \quad 9 = \begin{matrix} ||| \\ ||| \\ ||| \end{matrix}.$$

Tens, hundreds, and so on, were treated likewise. For example,

$$50 = \begin{matrix} \cap \cap \\ \cap \cap \cap \end{matrix}$$

$$700 = \begin{matrix} \mathcal{QQQ} \\ \mathcal{QQQQ} \end{matrix}.$$

These symbols were often combined to represent other numbers. For instance,

$$324 = \mathcal{QQQ} \cap \cap \; ||||.$$

Here the hundreds are represented first, then the tens, and finally the units, just as in modern notation. Hieroglyphics were also written from right to left, in which case the symbols themselves were reversed. For example, 324 could also be written as

$$|||| \cap \cap \; 999.$$

We further observe the following:

1. *A symbol for zero was lacking.* For instance, when writing 305, which we could not do without the zero, the Egyptian wrote

$$\mathcal{QQQ} \; \begin{matrix} || \\ ||| \end{matrix}.$$

2. *The numerals were written in base ten.* One symbol replaced 10 symbols of the next smaller denomination.

EXERCISES 1-3

1 See Figure 1-1. Determine the number of oxen, goats, and prisoners claimed by Menes on his mace. Compare your answers with those given in Section 1-1.

2 Write these numbers in hieroglyphics:
   (a) 53   (b) 407
   (c) 2136   (d) 12,345

3 What numbers are represented by the following:

(a) 𓂭𓂭 𓍢𓍢 |||
𓍢𓍢𓍢 ||||

(b) 𓁨𓁨

(c) 𓆐

4 How many different types of symbols are needed to write the numbers 1 through 1,000,000 in hieroglyphics? How many in our own numeration system?

5 Add, in hieroglyphics,

𓆼𓆼𓆼 𓍢𓍢𓍢 ||
𓆼𓆼𓆼 𓍢𓍢𓍢 ||| and 𓍢𓍢 |||
𓆼𓆼𓆼 ||| 𓍢𓍢𓍢 |||| .

How many number combinations did an Egyptian student need to memorize to be able to add? How many does a modern student need to memorize?

6 Multiply

𓂭 𓆼𓆼 ∩∩ ∩∩ ||| by ∩.

Can you suggest a simple rule for multiplying by 10 using Egyptian numerals?

1-4 ARITHMETIC OPERATIONS

In hieroglyphic notation, addition did not cause any difficulty. It was even simpler than in our system. There were no combinations such as $7 + 5 = 12$ to memorize. Since the Egyptians knew that 10 unit strokes could be replaced by ∩, 10 symbols ∩ by 𓍢, and so on, they could proceed by counting symbols in the two numbers to be added. Thus, they would write the sum of

𓍢𓍢𓍢 ∩∩ ||| and 𓍢𓍢 ||
𓍢𓍢𓍢 ∩∩ |||| 𓍢𓍢 |||

directly as

𓆼 ∩∩ ∩∩∩ || .

Having counted 10 vertical strokes, they wrote ∩ and then marked down the remaining two strokes without having to know that 7 plus 5 is 12 and without having to think: "I'll write the 2 and carry 1 (or 10) in my mind." And so on.

Subtraction was performed as shown by the following example. If Egyptians wanted to compute $12 - 5$, they thought: What will be needed to complete 5 to make 12? Such a completion was called *skm* (pronounced: saykam). We use a modern equivalent of this process in making change today. For instance, when $5.83 is paid with a $10 bill, the clerk counts the change from $5.83 up to $10.00; thus: $5.83 + $0.02 = $5.85; $5.85 + $0.05 = $5.90; $5.90 + $0.10 = $6.00; $6.00 + $4.00 = $10.00. The clerk does not say all of this as he counts the change into your hand, nor does he go back and add all the under-scored numbers—0.02, 0.05, 0.10, 4.00—to find the total amount of your change and hence the difference between $10.00 and $5.83. This completion process is mathematically sound. In ordinary algebra, and even in more advanced mathematical systems, as well as in arithmetic, subtraction is always the inverse of addition. Every subtraction problem, such as $12 - 5 = ?$, really gives the result (sum) of an addition and one of the addends and asks for the other addend. Thus, $12 - 5 = ?$ really means $12 = 5 + ?$. Mathematically, addition is a fundamental operation. Subtraction is defined in terms of addition and cannot exist without it. This fact is recognized when children are taught to check subtraction by addition, and when the subtraction facts are taught along with the addition facts.

The Egyptian method of multiplication was quite different from ours. The Egyptians used two operations to multiply: *doubling* and *adding*. To compute 6×8, for instance, they reasoned as follows:

$$2 \cdot 8 = 16$$
$$4 \cdot 8 = 2 \cdot (2 \cdot 8) = 32.$$

Addition on the left gives: $(2 + 4) \cdot 8$, or 6×8, and on the right: $16 + 32 = 48$. Hence, $6 \times 8 = 48$.

Problem 32 of the Rhind papyrus shows the actual procedure used by the Egyptians to compute 12×12. It goes as follows (reading from right to left):

It corresponds to the following (reading from left to right):

	1	12	
	2	24	
\	4	48	
\	8	96	sum 144.

From top to bottom we see the results of 1×12, 2×12, 4×12, and 8×12, which have been obtained by doubling. The sloping strokes next to the third and the fourth line indicate that only these lines are to be added to get the desired product. The symbol ⌐⊻┐ in the bottom line of the calculation in hieroglyphics represents a papyrus roll and means "the result is the following."

By way of an exception, the Egyptian sometimes multiplied a number directly by 10 instead of adding twice the number and eight times the number. This was easily done in his notation; he just substituted ∩ for I, ℰ for ∩, and so on.

EXAMPLE 1   Compute 14 × 80.

Hieroglyphics				Translation		
∩∩∩∩ ∩∩∩∩	I			1	80	
9999 9999	∩ /		\	10	800	
∩∩∩ ∩∩∩ 9	II			2	160	
∩∩9⅔ ⌐⊻┐ ∩∩999	IIII /	\		4	320	sum 1120.

Other approaches to multiplication were also used. For example, to multiply by 5, the Egyptian occasionally started by multiplying by 10 and then divided by 2.

EXAMPLE 2   Compute 16 × 16 (Kahun papyrus, Problem 6).

\ 1	16	
\ 10	160	
\ 5	80	sum 256.

Halving a number was considered to be a fundamental arithmetic operation that was done mentally.

The method of multiplication described in Example 2 was in use into the Hellenistic period, and as late as the Middle Ages doubling and halving were encountered as separate operations. In fact, under the heading of *duplation* (or *duplication*) and *mediation* they can be found as separate chapters in early American textbooks.

The reader will get a better understanding of Egyptian multiplication by trying his own hand at some computations.

EXERCISES 1-4

1 Write the following addition problems in hieroglyphics and then perform the addition.

(a) 46
   23

(b) 64
   28

(c) 4297
   1351

2 Repeat Exercise 1, with the operation changed to subtraction.

3 Without translating into hieroglyphics, compute by repeated doubling and adding:
   (a) 22 X 17   (b) 34 X 27   (c) 19 X 28

4 Write the computations of Exercise 3 in hieroglyphics.

5 (a) Write 426 in hieroglyphics.
   (b) Multiply 426 by 10, by changing the symbols.
   (c) Multiply 426 by 5, halving the number of each kind of symbol obtained in (b). Check your results by translating the symbols back into modern notation.

1-5 MULTIPLICATION

Look again at the multiplication 12 X 12:

$$\begin{array}{rl} 1 & 12 \\ 2 & 24 \\ \backslash \ 4 & 48 \\ \backslash \ 8 & 96 \qquad \text{sum } 144. \end{array}$$

The left-hand column consists of numbers that are powers of 2. The strokes indicate how 12, the multiplier, can be written as a sum of such powers.

The question might arise: Does this method of doubling and adding always succeed? It does, *if we can always write the multiplier as a sum of powers of 2.* Is this always possible? *The answer is:* Yes! An example will illustrate this. Suppose that we want to multiply 237 and 18, using 237 as the multiplier. This number would have to be written as a sum of powers of 2. The first nine powers of 2 are:

$$2^0 = 1 \qquad 2^3 = 8 \qquad 2^6 = 64$$
$$2^1 = 2 \qquad 2^4 = 16 \qquad 2^7 = 128$$
$$2^2 = 4 \qquad 2^5 = 32 \qquad 2^8 = 256.$$

Now let us write 237 as a sum of powers of 2:

$$237 = \underline{128} + 109,$$
$$\text{but } 109 = \underline{64} + 45,$$

$$\text{and } 45 = \underline{32} + 13,$$
$$\text{and } 13 = \underline{8} + 5,$$
$$\text{and } 5 = \underline{4} + \underline{1}.$$

Hence, $$237 = 128 + 64 + 32 + 8 + 4 + 1,$$

or $$237 = 2^7 + 2^6 + 2^5 + 2^3 + 2^2 + 2^0.$$

Every other multiplier can be written in a similar way as a sum of powers of 2.

Proceeding with the multiplication of 237 and 18, we get

$$237 \times 18 = (2^7 + 2^6 + 2^5 + 2^3 + 2^2 + 1) \times 18$$
$$= (2^7 \cdot 18) + (2^6 \cdot 18) + (2^5 \cdot 18)$$
$$+ (2^3 \cdot 18) + (2^2 \cdot 18) + (1 \cdot 18).$$

The Egyptian format for this computation would have been:

\	1	18
	2	36
\	4	72
\	8	144
	16	288
\	32	576
\	64	1152
\	128	2304
	237	4266.

In each column, only the numbers on the lines marked "\" are to be added.

In the preceding paragraph we used an extension of the distributive property, $(a + b) \cdot c = ac + bc$. This property has been used implicitly for centuries, and in this sense it is very old. However, only within the last century has the distributive property been recognized as a common fundamental principle, occurring as a part of the basic structure of a number of different mathematical systems. This recognition of the importance of underlying structures has become a characteristic of modern mathematics as well as a goal of mathematical research. Recognition of structure clarifies one's perceptions of old systems and may be a tool useful in inventing new ones. The distributive property is one of the defining properties of what is nowadays called a *field*. The reader is referred to Section 8-11 for a discussion of the concept of a field.

There is an exact parallel to the Egyptian multiplication in the

Russian peasant method of multiplication, said to be still in use today in some parts of Russia. In this method, all multiplications are performed by a combination of doubling and halving. Suppose that the Russian peasant wishes to multiply 154 by 83. He does this in a number of steps, each of which consists in halving one factor and doubling the other:

83 × 154   /

41 × 308   /   (41 is the "smaller half" of 83; 308 = 2 × 154)

20 × 616   (20 is the "smaller half" of 41)

10 × 1232   (10 is one half of 20)

5 × 2464   /

2 × 4928

1 × 9856   /

Cross off the lines containing an even number in the left-hand column. The lines with the stroke behind them will remain. Add the numbers remaining in the right-hand column, as follows:

$$154 + 308 + 2464 + 9856 = 12{,}782.$$

The required result is 12,782, which can be checked by ordinary multiplication.

Anybody seeing this procedure for the first time would naturally wonder whether it is correct, and if so, why? The explanation is as follows:

$83 = 82 + 1$ (so $83 \cdot 154 = 82 \cdot 154 + 1 \cdot 154$)

$82 = 41 \cdot 2$ (so $83 \cdot 154 = 41 \cdot 308 + \underline{154}$)

$41 = 40 + 1$ (so $83 \cdot 154 = 40 \cdot 308 + \underline{308} + \underline{154}$)

$40 = 20 \cdot 2$ (so $83 \cdot 154 = 20 \cdot 616 + \underline{308} + \underline{154}$)

$20 = 10 \cdot 2$ (so $83 \cdot 154 = 10 \cdot 1232 + \underline{308} + \underline{154}$)

$10 = 5 \cdot 2$ (so $83 \cdot 154 = 5 \cdot 2464 + \underline{308} + \underline{154}$)

$5 = 4 + 1$ (so $83 \cdot 154 = 4 \cdot 2464 + \underline{2464} + \underline{308} + \underline{154}$)

$4 = 2 \cdot 2$ (so $83 \cdot 154 = 2 \cdot 4928 + \underline{2464} + \underline{308} + 154$

and $83 \cdot 154 = \underline{9856} + \underline{2464} + \underline{308} + \underline{154}$),

or, finally,

$$83 \times 154 = 12{,}782.$$

A closer look at the left-hand column of our explanation of the Russian peasant multiplication will clarify its connection with Egyptian multiplication. The statements in that column could be combined to read thus:

$$83 = 82 + 1$$
$$= 41 \cdot 2 + 1$$
$$= (40 + 1) \cdot 2 + 1$$
$$= 40 \cdot 2 + 2 + 1$$
$$= 20 \cdot 2^2 + 2 + 1$$
$$= 10 \cdot 2^3 + 2 + 1$$
$$= 5 \cdot 2^4 + 2 + 1$$
$$= (4 + 1) \cdot 2^4 + 2 + 1$$
$$= 4 \cdot 2^4 + 1 \cdot 2^4 + 2 + 1$$
$$= 2^6 + 2^4 + 2 + 1.$$

This shows how 83 can be written as a sum of powers of 2 (remember, $1 = 2^0$). The same process can be applied to any counting number whatsoever.

Let us now compare the Russian peasant method with the Egyptian process for multiplying 83 and 154. The Egyptian way is:

\	1	154	
\	2	308	
	4	616	
	8	1232	
\	16	2464	
	32	4928	
\	64	9856	sum 12,782.

The strokes at the left indicate the powers of 2, which the Egyptian chose because their total is 83. The corresponding numbers in the right-hand column are: 154 $(= 154 \cdot 2^0)$, 308 $(= 154 \cdot 2^1)$, 2464 $(= 154 \cdot 2^4)$, 9856 $(= 154 \cdot 2^6)$. These are exactly the numbers marked off with strokes in the original Russian peasant multiplication (and underlined in our explanation of it).

Compared with our modern system, the ancient Eygptian multiplication is indeed a strange one. Division appears even more peculiar as done by the Egyptians. However, it is actually easier to understand than the algorithm most of us use. Instead of saying: "Calculate $45 \div 9$," an Egyptian said: "Calculate with 9 until 45 is reached." We start multiplying 9, as follows:

$$\begin{array}{ccl} \backslash & 1 & 9 \\ & 2 & 18 \\ \backslash & 4 & 36 \qquad \text{sum } 45. \end{array}$$

From this it then follows that $(1 + 4) \cdot 9 = 45$, or $45 \div 9 = 5$.

Today we often define division to be the inverse operation of multiplication. In other words, in every division problem we are given a product and one of its factors. The problem is to find the other factor. Thus, even today we teach that $45 \div 9 = ?$ means $? \times 9 = 45$.

EXERCISES 1-5

1 Compute the following by using ancient Egyptian multiplication:
   (a) 74×64 (b) 129×413
   (c) 58×692 (d) 4968×1234

2 Repeat Exercise 1, using the Russian peasant method.

3 Compute the product of Exercise 1(c) with the factors reversed. Does the order of the factors make a difference in the effort required? Explain.

4 Write the calculations of Exercises 1(a) and 1(b) in hieroglyphic symbols.

5 Compute the following by using ancient Egyptian division:
   (a) $360 \div 24$ (b) $238 \div 17$
   (c) $242 \div 11$ (d) $405 \div 9$

6 Represent the following numbers as sums of powers of 2:
   (a) 15   (b) 14
   (c) 22   (d) 45
   (e) 16   (f) 79
   (g) 968   (h) 8643

7 See the properties of a field stated in Section 8-11. Identify the properties that were used implicitly by the ancient Egyptians in
   (a) Example 1 of section 1-4
   (b) the multiplication 237×18 on page 12

1-6 FRACTIONS AND DIVISION

If the remainder of a division was not zero, fractions were introduced. Fractions were also used in the Egyptian system of weights and measures.

There is a marked difference between the Egyptian fractions and those we use. Our system admits any number as a numerator, whereas the Egyptians used only fractions with the numerator 1, with the exceptions $\frac{2}{3}$ and $\frac{3}{4}$. We shall call fractions with numerator 1 *unit fractions*.

To write such a fraction, the Egyptians merely wrote the denominator below the symbol $\bigcirc$, which symbol represented an open mouth. For example, $\frac{1}{12}$ is $\bigcirc\!\!\!\!\!\underset{\cap II}{}$ in hieroglyphics.

The Egyptian had separate symbols for a few fractions. Some of these were

$$\frac{1}{2} = \diagup\!\!\!\!\!\diagdown \ , \qquad \frac{1}{4} = \times , \qquad \frac{2}{3} = \maltese .$$

There was also a symbol for $\frac{3}{4}$, $\maltese$. However, in later writings $\frac{3}{4}$ was written as

$$\diagup\!\!\!\!\!\diagdown \ \times \ (= \frac{1}{2} + \frac{1}{4}).$$

The use of special symbols for frequently occurring fractions is similar to the use in English of words such as "one half," "one quarter," "one percent" for special common fractions rather than "one twoth," "one fourth," "one one-hundredth."

In working out computations, the Egyptian often came across results that could not be expressed by single unit fractions. In that case he wrote them as a sum of different unit fractions. For example, an Egyptian might write $\frac{1}{3} + \frac{1}{2}$ for $\frac{5}{6}$. We have already seen the reduction $\frac{3}{4} = \frac{1}{2} + \frac{1}{4}$.

Let us immediately remark that such reductions are not always uniquely determined. Consider these examples:

EXAMPLE 1

$$\frac{7}{24} = \frac{4+3}{24} = \frac{4}{24} + \frac{3}{24} = \frac{1}{6} + \frac{1}{8} ,$$

but also
$$\frac{7}{24} = \frac{6}{24} + \frac{1}{24} = \frac{1}{4} + \frac{1}{24} .$$

EXAMPLE 2

$$\frac{2}{35} = \frac{6}{105} = \frac{1}{21} + \frac{1}{105} ,$$

but also
$$\frac{2}{35} = \frac{12}{210} = \frac{1}{30} + \frac{1}{42} ,$$

or
$$\frac{2}{35} = \frac{8}{140} = \frac{1}{20} + \frac{1}{140} .$$

The unsettled question of how the Egyptians found their unit fraction representations has stimulated several mathematicians to study this problem. *J. J. Sylvester* (1814-1897) proposed a system for expressing in a unique way every fraction between zero and 1 as a sum of unit fractions. His process calls for (1) finding the largest unit fraction (the one with the smallest denominator) less than the given fraction, (2) subtracting this unit fraction from the given fraction, (3) finding the largest unit fraction less than the resulting difference, (4) subtracting again, and continuing this process. Let us apply Sylvester's process to the fractions considered in Examples 1 and 2.

1. Of the unit fractions $\frac{1}{2}, \frac{1}{3}, \frac{1}{4}, \frac{1}{5}, \ldots$, the fraction $\frac{1}{4}$ is the largest unit fraction that is less than $\frac{7}{24}$. Subtraction gives

$$\frac{7}{24} - \frac{1}{4} = \frac{1}{24},$$

from which it follows that

$$\frac{7}{24} = \frac{1}{4} + \frac{1}{24},$$

in agreement with one of the results of Example 1.

2. The largest unit fraction less than $\frac{2}{35}$ is $\frac{1}{18}$. Subtraction gives

$$\frac{2}{35} - \frac{1}{18} = \frac{1}{630},$$

from which it follows that

$$\frac{2}{35} = \frac{1}{18} + \frac{1}{630}.$$

This outcome differs from the results in Example 2.

In the following example, Sylvester's method leads to a sum of more than two unit fractions.

EXAMPLE 3   Find a unit fraction representation for $\frac{13}{20}$.

The largest unit fraction less than $\frac{13}{20}$ is $\frac{1}{2}$; thus

$$\frac{13}{20} - \frac{1}{2} = \frac{3}{20}.$$

The largest unit fraction less than $\frac{3}{20}$ is $\frac{1}{7}$; thus,

$$\frac{3}{20} - \frac{1}{7} = \frac{1}{140}.$$

Hence,
$$\frac{13}{20} = \frac{1}{2} + \frac{1}{7} + \frac{1}{140}.$$

Sylvester not only invented the procedure but he also proved that in this way every fraction can be represented as a finite sum of unit fractions.

In order to follow ancient Egyptian processes more easily, we shall use a new notation for unit fractions. The fraction $\frac{1}{12}$, for instance, will be represented by $\overline{12}$, and, in general, $\frac{1}{n}$ by $\overline{n}$. The fraction $\frac{2}{3}$ will be written $\overline{\overline{3}}$.

The following examples show how the Egyptians did divisions that do not have a whole-number quotient.

EXAMPLE 4   (Problem 24 of the Rhind papyrus.) Divide 19 by 8 (calculate with 8 until you find 19).

$$
\begin{array}{ccl}
 & 1 & 8 \\
\backslash & 2 & 16 \\
 & \overline{2} & 4 \\
\backslash & \overline{4} & 2 \\
\backslash & \overline{8} & 1 \qquad \text{sum 19.}
\end{array}
$$

The total of the numbers on the marked lines in the right-hand column is 19. Evidently,

$$2 \times 8 + \overline{4} \times 8 + \overline{8} \times 8 = (2 + \overline{4} + \overline{8}) \times 8 = 19;$$

hence,
$$19 \div 8 = 2 + \overline{4} + \overline{8},$$

which can easily be checked by using conventional notation and multiplication.

In addition to performing a division by multiplying consecutively

by
$$\overline{2}, \overline{4}, \overline{8}, \ldots,$$

the Egyptians used the sequence

$$\overline{\overline{3}}, \overline{3}, \overline{6}, \ldots$$

when it was more convenient. In this case, it is remarkable that they first multiplied by two thirds, and then by one third.

EXAMPLE 5 Compute $20 \div 24$ (calculate with 24 until you find 20).

	1	24
\	$\overline{\overline{3}}$	16	(since this is less than 20, put a stroke before it; we need 4 more, so continue)
	$\overline{3}$	8	(since this is greater than 4, there is no stroke)
\	$\overline{6}$	4	sum 20

Hence, $20 \div 24 = \overline{\overline{3}} + \overline{6} \left(= \dfrac{2}{3} + \dfrac{1}{6} \right)$.

In the preceding examples and exercises, the divisions could be performed with one of the two series

$$\overline{2}, \overline{4}, \overline{8}, \overline{16}, \ldots$$

or

$$\overline{\overline{3}}, \overline{3}, \overline{6}, \overline{12}, \ldots$$

However, not every division problem can be solved by using only halves and thirds. For that reason other methods were sometimes used.

EXAMPLE 6 Compute $11 \div 15$.

	1	15
\	$\overline{\overline{3}}$	10
\	$\overline{15}$	1	sum 11;

hence, $11 \div 15 = \overline{\overline{3}} + \overline{15}.$

After the second line we might have expected

$\overline{3}$	5
$\overline{6}$	2 $\overline{2}$
etc.	

However, we find

$$\overline{15} \quad 1.$$

Apparently, the reasoning was as follows:

	1	15
	$\overline{\overline{3}}$	10	(note that this is nearly 11, so we only need 1 more; since 1 is one fifteenth of 15, we continue)
	$\overline{15}$	1.	

EXAMPLE 7   Compute $9 \div 24$.

$$
\begin{array}{rl}
& \overline{\overline{1}} \quad 24 \\
& \overline{3} \quad 16 \\
\backslash \;\; & \overline{3} \quad 8 \\
\backslash \;\; & \overline{24} \quad 1.
\end{array}
$$

(note that we are near the end; we need only 1 more, so)

Hence, $9 \div 24 = \overline{3} + \overline{24}.$

From the methods of multiplication and division described above, it appears that Egyptian arithmetic was essentially additive. That is, the chief arithmetical operation was addition. Subtraction was reduced to addition. Multiplication was done by doubling and adding. Division was done by halving or doubling, and then adding.

EXERCISES 1-6

1 Compute in the ancient Egyptian way:
   (a) $26 \div 20$ (b) $55 \div 6$
   (c) $71 \div 21$ (d) $25 \div 18$
   (e) $52 \div 68$ (f) $13 \div 36$

2 Compute in the ancient Egyptian way:
   (a) $3 \div 4$ (b) $5 \div 8$
   (c) $14 \div 24$ (d) $35 \div 32$
   (e) $5 \div 6$ (f) $17 \div 12$
   (g) $11 \div 16$ (h) $51 \div 18$

3 Find the Sylvester-type representation (as a sum of unit fractions) for each of the following:
   (a) $\dfrac{13}{36}$ (b) $\dfrac{9}{20}$
   (c) $\dfrac{4}{15}$ (d) $\dfrac{335}{336}$

   (Hint: To find the largest unit fraction less than the given fraction, divide the denominator by the numerator and take the next integer greater than the quotient for the new denominator.)

4 Find a non-Sylvester-type representation for the fractions in (a), (b), and (c) of Exercise 3. (Hint: See Examples 1 and 2 of this section.)

5 There are various procedures for finding equivalent sums of unit fractions in special cases.
   (a) If m is an odd number, then

$$
\frac{2}{m} = \frac{1}{m \cdot \dfrac{m+1}{2}} + \frac{1}{\dfrac{m+1}{2}} .
$$

   Verify that this is true for m equal to $3, 5,$ and 7.

(b) Prove the theorem in (a).

(c) Why is there no need for a corresponding theorem for m an *even* number?

6 (a) Prove that if n is an integer, then

$$\frac{1}{n} = \frac{1}{n+1} + \frac{1}{n(n+1)}.$$

(b) Use the identity in (a) to write $\frac{1}{3}$ and $\frac{1}{4}$ as sums of unit fractions.

(c) Use the identity in (a) to prove: If a rational number can be represented as a sum of unit fractions in one way, then this number can be represented as a sum of unit fractions in an infinite number of ways.

7 Prove that every rational number can be represented as the sum of a finite number of unit fractions. (Hint: Apply Sylvester's process, using the hint of Exercise 3.)

8 Prove that Sylvester's process leads to a unique representation for each rational number.

9 There is a similarity between Sylvester's process and the expansion of both rational and irrational numbers into continued fractions. See reference 8 or 9 at the end of the chapter, for example, for a discussion of continued fractions and find continued-fraction expansions for the numbers in Exercise 3.

10 See the properties of a field stated in Section 8-11. Identify the properties that were used implicitly by the ancient Egyptians in
(a) Example 4 of this section,
(b) Example 5 of this section.

1-7 THE RED AUXILIARY NUMBERS

The sequences $\overline{2}, \overline{4}, \ldots$ and $\overline{\overline{3}}, \overline{3}, \overline{6}, \ldots$ do not always lead to a representation of a quotient as a sum of unit fractions in a finite number of steps. It became necessary to invent a new trick. This is demonstrated in the following example.

EXAMPLE 1 Compute $5 \div 17$.

Unsuspectingly we start:

$$
\begin{array}{ccc}
1 & 17 & \\
\overline{2} & 8 & \overline{2} \\
\backslash\ \overline{4} & 4 & \overline{4}.
\end{array}
\qquad \left(\text{think: } 8 + \frac{1}{2}\right)
$$

We wish to obtain numbers in the right-hand column that will add up to 5. If we continue normally, we get

$$
\begin{array}{ccc}
\overline{8} & 2 & \overline{8} \\
\overline{16} & 1 & \overline{16},
\end{array}
$$

etc.

But now we are getting into a hopeless muddle, for on the right-hand side there are fractions with denominators that increase continually, and the chance that any combinations of these fractions will add up to 5 becomes extremely slight, if it is possible at all. For this reason, the process changes after the third line. Beginning again,

$$
\begin{array}{rcl}
1 & 17 & \\
\overline{2} & 8 & \overline{2} \\
\backslash \quad \overline{4} & 4 & \overline{4}.
\end{array}
$$

The Egyptians may have reasoned as follows. We have nearly reached the end. We wish to obtain 5 on the right-hand side. Hence, we are still $\overline{2} + \overline{4}$ short. (How can you tell?) What must we have in the left-hand column to get $\overline{2} + \overline{4}$ in the right-hand one? Well, first multiply 17 by $\overline{17}$ to get 1. So, once more, and now for the last time:

$$
\begin{array}{rcll}
1 & 17 & & \\
\overline{2} & 8 & \overline{2} & \\
\backslash \quad \overline{4} & 4 & \overline{4} & \\
\overline{17} & 1 & & \\
\backslash \quad \overline{34} & \overline{2} & & \\
\backslash \quad \overline{68} & \overline{4} & & \text{sum } 4 + \overline{4} + \overline{2} + \overline{4} = 5.
\end{array}
$$

Hence, $5 \div 17 = \overline{4} + \overline{34} + \overline{68}$.

The problem that arises after the third line is: How do we "complete" $4 + \overline{4}$ to 5? or: How do we complete $\overline{4}$ to 1? The answer is to use a *skm* (or subtraction) computation, such as was discussed in Section 1-4. Such a *skm* is not always as simple as in this example. That is why in the Rhind papyrus considerable attention was given to this kind of computation. Strikingly enough, some numbers were written in red to make them immediately conspicuous. The numbers that appeared in red are printed in boldface type in the following examples.

EXAMPLE 2   How is $\overline{15} + \overline{3} + \overline{5}$ completed to 1?

The calculation was as follows:

$$
\overline{15} + \overline{3} + \overline{5}
$$
$$
\mathbf{1} \quad \mathbf{5} \quad \mathbf{3} \qquad \text{sum 9, remainder 6;}
$$

calculate with 15 until you find 6.

$$
\begin{array}{cc}
1 & 15 \\
\overline{\overline{3}} & 10 \\
\backslash \ \ \overline{3} & 5 \\
\backslash \ \ \overline{15} & 1 \qquad \text{sum 6.}
\end{array}
$$

The answer is $\qquad \overline{15} + \overline{3}$.

Apparently, $(\overline{15} + \overline{3} + \overline{5}) + (\overline{15} + \overline{3}) = 1$, which can be easily verified by the reader.

At first sight, this method is quite unintelligible. What actually happened?

Let us tackle this problem with our modern arithmetic. We state the problem as

$$\text{complete the sum } \frac{1}{15} + \frac{1}{3} + \frac{1}{5} \text{ to } 1.$$

We then change the fractions to the same denominator:

$$\frac{1}{15} + \frac{1}{3} + \frac{1}{5} = \frac{1}{15} + \frac{5}{15} + \frac{3}{15} = \frac{9}{15}.$$

We still need $\frac{6}{15}$. Since the Egyptians did not have a symbol for $\frac{6}{15}$ as such, they had to find a set of unit fractions whose sum is $\frac{6}{15}$. Therefore, they calculated with 15 until they found 6. That is, they divided 6 by 15. But what is the meaning of the red auxiliary numbers? They are the numerators of the given fractions after these fractions have been written with the common denominator 15:

$$\frac{1}{15}, \ \frac{5}{15}, \ \frac{3}{15}.$$

Should the red auxiliary numbers actually be regarded as numerators? Perhaps, but an objection is that the red auxiliary numbers are sometimes fractions, and fractions as numerators of fractions are, to understate the case, inconvenient. It is therefore more probable that the red auxiliary numbers have to be explained as follows.

Choose a new unit that is $\frac{1}{15}$ as large as the old one.

$$
\begin{array}{lll}
1 & \text{is now called} & 15, \\
\overline{15} & \text{is now called} & 1, \\
\overline{3} & \text{is now called} & 5, \\
\overline{5} & \text{is now called} & 3.
\end{array}
$$

The previous $\overline{15} + \overline{3} + \overline{5}$ now corresponds to $1 + 5 + 3$, and the problem of completing $\overline{15} + \overline{3} + \overline{5}$ to 1 now corresponds to the problem of completing $1 + 5 + 3$ to 15. So 6 new units are to be added. But a new unit is $\overline{15}$ of the old unit. Hence, the 6 new units are $6 \div 15$ old units. Calculate with 15 until you find 6; this gives the quotient $\overline{15} + \overline{3}$. This is the required completion.

We shall now perform more intricate divisions using *skm* computations. For example, let us compute $11 \div 25$.

	1	25	
	$\overline{\overline{3}}$	16	$\overline{\overline{3}}$
\	$\overline{3}$	8	$\overline{3}$
	$\overline{6}$	4	$\overline{6}$
\	$\overline{12}$	2	$\overline{12}.$

On the right-hand side, the marked lines give the sum $10 + \overline{3} + \overline{12}$. We need 11. What is the difference between $10 + \overline{3} + \overline{12}$ and 11? This amounts to the question: How is $\overline{3} + \overline{12}$ completed to 1? The Egyptians chose a new unit, which was $\frac{1}{12}$ of the old unit. Then the original unit, 1, would be represented by 12. This number had the advantage that it was easy to find both $\frac{1}{3}$ and $\frac{1}{12}$ of it. So the Egyptians continued:

$\overline{3}$	$\overline{12}$	
4	1	sum 5, remainder 7.

(That is, the Egyptian thought process seems to have been: $\frac{1}{3}$ of 12 is 4; $\frac{1}{12}$ of 12 is 1; $4 + 1 = 5$; $12 - 5 = 7$. Hence, the required difference between 1 and $\overline{3} + \overline{12}$ is 7 new units and is now going to be calculated in terms of the old unit.)

Calculate with 12 until you find 7.

	1	12	
\	$\overline{2}$	6	
\	$\overline{12}$	1	sum 7.

Hence, $\dfrac{7}{12} = \overline{2} + \overline{12}.$

But the end is not yet in sight. In the right-hand column of the original computation there must be the sum of $\overline{2} + \overline{12}$ that we just found or the numbers $\overline{2}$ and $\overline{12}$ separately. What, then, should the left-hand column show? Or: By what number should 25 be multiplied to get this $\overline{2} + \overline{12}$ (or $\overline{2}$ and $\overline{12}$ separately) in the right-hand column?

Now 25 is multiplied by $\overline{25}$ to get 1, and the continuation of the original computation reads:

$$\begin{array}{ccc} & \overline{25} \text{ (of 25)} & 1 \\ \backslash & \overline{50} & \overline{2} \\ \backslash & \overline{300} & \overline{12}. \end{array}$$

Hence, the whole calculation proceeds as follows:

$$\begin{array}{cccc} & 1 & 25 & \\ & \overline{\overline{3}} & 16 & \overline{\overline{3}} \\ \backslash & \overline{3} & 8 & \overline{3} \\ & \overline{6} & 4 & \overline{6} \\ \backslash & \overline{12} & 2 & \overline{12} \\ & \overline{25} & 1 & \\ \backslash & \overline{50} & \overline{2} & \\ \backslash & \overline{300} & \overline{12} & \quad \text{sum 11.} \end{array}$$

Hence, $\qquad 11 \div 25 = \overline{3} + \overline{12} + \overline{50} + \overline{300}.$

EXERCISES 1-7

1 Complete to 1, using a *skm* computation. Check your answers.
  (a) $\overline{12} + \overline{6} + \overline{3}$ (b) $\overline{5} + \overline{4} + \overline{3}$
  (c) $\overline{14} + \overline{4} + \overline{7}$ (d) $\overline{3} + \overline{6} + \overline{9}$
  (e) $\overline{3} + \overline{4}$ (f) $\overline{12} + \overline{9} + \overline{18}$
  (g) $\overline{4} + \overline{6} + \overline{8}$ (h) $\overline{5} + \overline{12} + \overline{15}$

2 Perform the following divisions in the ancient Egyptian way. Check your answers.
  (a) $12 \div 23$ (b) $11 \div 13$
  (c) $15 \div 19$ (d) $33 \div 7$
  (e) $11 \div 65$ (f) $9 \div 23$

3 Show that $\frac{5}{17}$ cannot be represented as the sum of a finite number of unit fractions whose denominators are powers of 2.

1-8 THE 2 ÷ n TABLE

Readers who think that they have now seen all there is to Egyptian division are mistaken. The Egyptians now proceeded to perform divisions in which the divisor as well as the dividend was a fractional number. To do this more quickly, they set up tables. In the Rhind papyrus, for example, we find a table for the reduction of fractions of the form $2 \div n$

to a sum of unit fractions (see Table 1-1). The numerator is always 2 and the denominators are the odd numbers from 3 through 101. There are no even denominators because $2 \div 12$, for instance, can immediately be replaced by $1 \div 6$, or $\overline{6}$, which is a unit fraction.

Table 1-1   Table of $2 \div n$

$2 \div 3 = \overline{2} + \overline{6}$	$2 \div 53 = \overline{30} + \overline{318} + \overline{795}$
$2 \div 5 = \overline{3} + \overline{15}$	$2 \div 55 = \overline{30} + \overline{330}$
$2 \div 7 = \overline{4} + \overline{28}$	$2 \div 57 = \overline{38} + \overline{114}$
$2 \div 9 = \overline{6} + \overline{18}$	$2 \div 59 = \overline{36} + \overline{236} + \overline{531}$
$2 \div 11 = \overline{6} + \overline{66}$	$2 \div 61 = \overline{40} + \overline{244} + \overline{488} + \overline{610}$
$2 \div 13 = \overline{8} + \overline{52} + \overline{104}$	$2 \div 63 = \overline{42} + \overline{126}$
$2 \div 15 = \overline{10} + \overline{30}$	$2 \div 65 = \overline{39} + \overline{195}$
$2 \div 17 = \overline{12} + \overline{51} + \overline{68}$	$2 \div 67 = \overline{40} + \overline{335} + \overline{536}$
$2 \div 19 = \overline{12} + \overline{76} + \overline{114}$	$2 \div 69 = \overline{46} + \overline{138}$
$2 \div 21 = \overline{14} + \overline{42}$	$2 \div 71 = \overline{40} + \overline{568} + \overline{710}$
$2 \div 23 = \overline{12} + \overline{276}$	$2 \div 73 = \overline{60} + \overline{219} + \overline{292} + \overline{365}$
$2 \div 25 = \overline{15} + \overline{75}$	$2 \div 75 = \overline{50} + \overline{150}$
$2 \div 27 = \overline{18} + \overline{54}$	$2 \div 77 = \overline{44} + \overline{308}$
$2 \div 29 = \overline{24} + \overline{58} + \overline{174} + \overline{232}$	$2 \div 79 = \overline{60} + \overline{237} + \overline{316} + \overline{790}$
$2 \div 31 = \overline{20} + \overline{124} + \overline{155}$	$2 \div 81 = \overline{54} + \overline{162}$
$2 \div 33 = \overline{22} + \overline{66}$	$2 \div 83 = \overline{60} + \overline{332} + \overline{415} + \overline{498}$
$2 \div 35 = \overline{30} + \overline{42}$	$2 \div 85 = \overline{51} + \overline{255}$
$2 \div 37 = \overline{24} + \overline{111} + \overline{296}$	$2 \div 87 = \overline{58} + \overline{174}$
$2 \div 39 = \overline{26} + \overline{78}$	$2 \div 89 = \overline{60} + \overline{356} + \overline{534} + \overline{890}$
$2 \div 41 = \overline{24} + \overline{246} + \overline{328}$	$2 \div 91 = \overline{70} + \overline{130}$
$2 \div 43 = \overline{42} + \overline{86} + \overline{129} + \overline{301}$	$2 \div 93 = \overline{62} + \overline{186}$
$2 \div 45 = \overline{30} + \overline{90}$	$2 \div 95 = \overline{60} + \overline{380} + \overline{570}$
$2 \div 47 = \overline{30} + \overline{141} + \overline{470}$	$2 \div 97 = \overline{56} + \overline{679} + \overline{776}$
$2 \div 49 = \overline{28} + \overline{196}$	$2 \div 99 = \overline{66} + \overline{198}$
$2 \div 51 = \overline{34} + \overline{102}$	$2 \div 101 = \overline{101} + \overline{202} + \overline{303} + \overline{606}$

The papyrus explains how some of the table entries were obtained. Some can be found in the manner explained previously. However, not all the partitions in the $2 \div n$ table fit these methods, and mathematicians have sought other explanations. The simplest partition of $2 \div n$ would have been $\overline{n} + \overline{n}$, but in writing a sum of unit fractions, the Egyptian never repeated a unit fraction. Another partition could be found by observing that $2 = 1 + \overline{2} + \overline{3} + \overline{6}$, and hence

$$2 \div n = \frac{2}{n} = \frac{1 + \overline{2} + \overline{3} + \overline{6}}{n} = \overline{n} + \overline{2n} + \overline{3n} + \overline{6n}.$$

However, such partitions are not to be found in the table, with the exception of $2 \div 101$. The Egyptians did know such partitions, but apparently they compiled the table so that they would have additional partitions which might be useful. We do not know why the partitions in the table were selected from those available. It does seem to be clear that the Egyptians did not want to use fractions with a denominator greater than 1000. The partition for $2 \div 101$ is the only possible one with all the denominators less than 1000 (if we do not consider $\overline{101} + \overline{101}$).

The following example shows how the $2 \div n$ table was used as a computational aid.

EXAMPLE 1   Divide 18 $\overline{4}$ $\overline{28}$ by 1 $\overline{7}$.

The computation proceeds thus:

1	1 $\overline{7}$	
2	2 $\overline{4}$ $\overline{28}$	(here we use the fact that $2 \times \overline{7} = 2 \div 7$ and that the table gives $2 \div 7 = \overline{4}$ $\overline{28}$)
4	4 $\overline{2}$ $\overline{14}$	
8	9 $\overline{7}$	
\ 16	18 $\overline{4}$ $\overline{28}$	(table used again).

Hence, the quotient is 16.

From this example it appears that the $2 \div n$ table was used in the doubling of fractions, a process needed in both multiplication and division. Notice that a *skm* computation was not needed in this example. However, the Egyptians did not always treat their pupils so mildly. They did not shrink from requiring intricate *skm* computations. As a forbidding example we give the following problem, to be found as Problem 33 in the Rhind papyrus:

Divide 37 by 1 $\overline{\overline{3}}$ $\overline{2}$ $\overline{7}$.

Readers who have the courage to try the direct approach will soon notice that it seems to lead away from a solution. There is nothing to be done but to perform a *skm* when 37 has nearly been reached. The result is 16 $\overline{56}$ $\overline{679}$ $\overline{776}$, which readers may check if they like!

Our discussion of ancient Egyptian computations shows that in spite of their poor notation, the Egyptians had attained great proficiency in the techniques of arithmetic. Considering the trouble that today's students sometimes have in computing with fractions, even in modern, simple notation, we must admire the patience and the acumen of these people, who knew how to work out such complicated problems 4000 years ago.

EXERCISES 1-8

1 Use ancient Egyptian division to calculate:
   (a) 4 $\overline{\overline{3}}$ $\overline{10}$ $\overline{30}$ ÷ 1 $\overline{5}$
   (b) 17 $\overline{3}$ $\overline{5}$ $\overline{15}$ ÷ 2 $\overline{5}$
2 Use a *skm* process to calculate 37 ÷ 1 $\overline{\overline{3}}$ $\overline{2}$ $\overline{7}$.

1-9 THE LEATHER ROLL

In Section 1-3 we became acquainted with hieroglyphics. To draw them took time, and therefore we need not be surprised that as Egyptians wrote more, they developed a simpler notation, *hieratic* writing. Hieroglyphics were used in inscriptions chiseled in stone. The cursive hieratic writing was used on papyri and also on the leather roll that we mentioned earlier. The contents of the latter are disappointing. They are merely a set of calculations that illustrate the addition of fractions. Some of them are very simple: for example, $\overline{10}$ + $\overline{10}$ = $\overline{5}$.

However, we include a picture of the leather roll (Figure 1-4) because its clearness and simplicity make it easy to decipher. Table 1-2 explains the hieratic signs in the leather roll.

Table 1-2

1	|	10	Λ	100	⁊	1000	ʃ		
2	||	20	Λ	200	⁊				
3	|||	30	Λ	300	⁊				
4	— or ||||	40	⸱	400	⁊		$\frac{1}{2}$	⊃
5	ᵞ		50	⼺	500	⁊		$\frac{1}{3}$	⸜
6	||| or Ƶ	60	⊔	600	⁊		$\frac{1}{4}$	X
7	2	70	⼺	700	⁊		$\frac{2}{3}$	⼡
8	=	80	⊔⊔	800	⁊			
9	⼈	90	⊔⊔	900	⁊			

A dot over a figure indicates a fraction. Thus, = Λ = 18 and = $\dot{\Lambda}$ = $\overline{18}$. The dot is placed over the symbol for the largest unit. The leather roll must be read from right to left.

To the left of the columns we find the sign ʃ ⊔⊔. It represents the demonstrative pronoun, and in this context we have to read it as "that is."

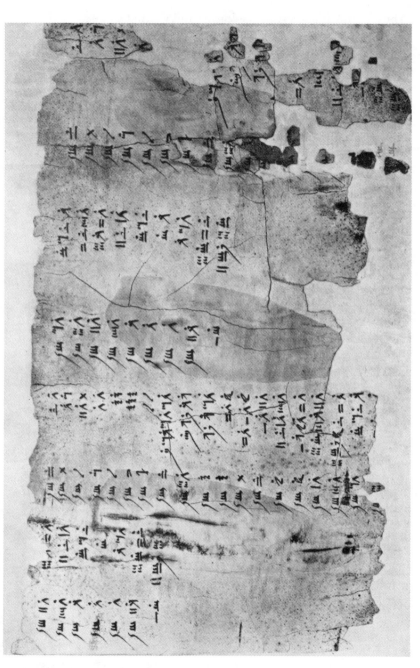

Figure 1-4 The leather roll (BM 10250). (From B. L. van der Waerden, *Science Awakening* 1, Wolters-Noordhoff Publishing, Groningen, The Netherlands, 1954.)

Thus, in the upper left of the leather roll we find

$$\int ⁛ \qquad || \dot{\wedge} \qquad ||| \dot{\wedge} \qquad = \dot{\wedge}$$

which means

"that is" $\overline{12}$ $\overline{36}$ $\overline{18}$.

This has to be read from right to left. So we would write

$$\frac{1}{18} + \frac{1}{36} = \frac{1}{12}.$$

It should not be difficult to continue reading the leather roll.

EXERCISES 1-9

1 Write in hieratic:
   (a) 1275   (b) 901
   (c) 91   (d) 910

2 Translate and check the problem on the second line of the two left-hand columns in Figure 1-4.

3 Translate and check the problem on the second line of the fifth and sixth columns in Figure 1-4.

4 What is the largest number that you find represented in Figure 1-4?

5 What is the smallest number that you find represented in Figure 1-4?

1-10   ALGEBRAIC PROBLEMS

Although most Egyptian mathematics was arithmetic, with applications to the measurement of geometric figures, we can see forerunners of several topics now included in the subject matter of high school algebra. This is particularly true of the *aha* problems. *Aha* means "heap," or "quantity." The first *aha* problem in the Rhind papyrus is Problem 24, which is translated in the steps that follow. (We have numbered the steps so that we can refer to them later.)

1. A quantity and its seventh, added together, give 19. What is the quantity?

2. Assume 7.   \ 1   7
                      \ $\overline{7}$ 1

                      Total   8.

As many times as 8 must be multiplied to give 19, so many times 7 will give the required number. (Note that we did this computation in Section 1-6.)

3.

	1	8
\	2	16
	$\overline{2}$	4
\	$\overline{4}$	2
\	$\overline{8}$	1
	Total	2 $\overline{4}$ $\overline{8}$

(that is, $19 \div 8 = 2 + \overline{4} + \overline{8}$).

4.

\	1	2	$\overline{4}$	$\overline{8}$
\	2	4	$\overline{2}$	$\overline{4}$
\	4	9	$\overline{2}$.	

5. Do it thus. The quantity is

16 $\overline{2}$ $\overline{8}$ (that is, $7 \times (2 + \overline{4} + \overline{8}) = 16 + \overline{2} + \overline{8}$, the solution)

2 $\overline{4}$ $\overline{8}$ (add this to 16 $\overline{2}$ $\overline{8}$)

Total   19   (the solution, $16 + \overline{2} + \overline{8}$, checks).

This becomes clearer when translated into a modern algebraic form.

$1'$. $x + \frac{1}{7}x = 19$

$2'$. Assume that $x = 7$; then $7 + \frac{1}{7} \cdot 7 = 8$ ($8 \neq 19$; hence step $3'$)

$3'$. $19 \div 8 = 2 + \frac{1}{4} + \frac{1}{8} = 2\frac{3}{8}$

$4'$. $7 \cdot (2\frac{3}{8}) = 16 + \frac{1}{2} + \frac{1}{8} = 16\frac{5}{8}$

$5'$. $16\frac{5}{8} + \frac{1}{7} \cdot (16\frac{5}{8}) = 16\frac{5}{8} + 2\frac{3}{8} = 19$

The 7 that was used as a replacement for *aha* in step 2 (and $2'$) was not thought of as the correct value, or even as an approximation to the correct value. On the other hand, although *aha*, or *x*, could have been replaced by any number, the Egyptians had a good reason for choosing 7 in this case rather than 8 or 9 or some other number. It was convenient to use 7, because $\frac{1}{7} \times 7 = 1$. The use of 7 as the assumed replacement avoided fractions in step 2.

The introduction of an arbitrary number at the beginning of a solution, such as the number 7 in the preceding *aha* problem, was used centuries later by Western European mathematicians. They called it the *method of single false position*. At the same time, a method for treating similar but more complicated problems, called the *method of double false position*, was developed. Both of these methods were taught to American students as recently as the latter part of the nineteenth century (see Exercise 3 of Exercises 1-10). Probably none of these users realized that the Egyptians had applied the same method centuries earlier.

There is only one problem known where the Egyptians used the method of single false position on a nonlinear equation, the first problem in the Berlin papyrus. The problem is as follows: "The sum of the areas of two squares is 100. Three times the side of one is four times the side of the other. Find the sides of the squares."

In modern notation, this problem would appear as a pair of equations in two unknowns:

$$x^2 + y^2 = 100$$
$$3x = 4y.$$

The Egyptian process of solving these equations was (in modern notation):

1. Take $x = 4$; then $y = 3$
2. Then $4^2 + 3^2 = 25$ $(25 \neq 100$; hence step 3$)$
3. $\sqrt{25} = 5$, $\sqrt{100} = 10$
4. $10 \div 5 = 2$
5. The sides are $2 \times 3 = 6$ and $2 \times 4 = 8$

This method can be shown to work for all equations of this type, that is, those equations in two unknowns in which the terms containing the unknowns are all of the same degree.

EXERCISES 1-10

1 Use the method of false position to solve the following problems. Then check your work by using an equation.

   (a) Rhind papyrus Problem 25: "A quantity and its $\frac{1}{2}$ added together become 16. What is the quantity?"
   (b) Rhind papyrus Problem 28: "A quantity and its $\frac{2}{3}$ are added together and from the sum $\frac{1}{3}$ of the sum is subtracted, and 10 remains. What is the quantity?"

2 Problem 40 of the Rhind papyrus involves the idea of an arithmetic progression and is similar to problems of a type that appeared centuries later under the heading "partnership" or "inheritance." The problem is as follows: "Divide 100 loaves among five men in such a way that the shares received shall be in arithmetical progression and that $\frac{1}{7}$ of the sum of the largest three shares shall be equal to the sum of the smallest two. What is the difference of the shares?" The papyrus uses false position in this, too.

   (a) Find the shares by whatever modern method you choose.
   (b) Use false position to find them.

3 Figures 1-5(a) through 1-5(e) are from *Daboll's Schoolmaster's Assistant*, first
   published in the United States in 1800 and until 1850 the most popular arithmetic textbook in the country. It was a revision by David Daboll of an English
   textbook by Richard Daboll. The book's method follows the teaching of that
   day. The students were given a rule, a worked example, and then examples to
   do themselves. Very little or no explanation was given.
   (a) Study the explanation under "Position" and "Single Position" in Figures
       1-5(b) and 1-5(c). Then work examples 2 and 3.
   (b) Study the explanation under "Double Position" in Figures 1-5(d) and 1-5(e).
       'Then work examples 2 and 3.
   (c) Work the remaining examples in Figures 1-5(c), 1-5(d), and 1-5(e).
   (d) Do the examples in (a) and (b) by using modern algebraic procedures.

4 Another popular early arithmetic textbook, *The American Tutor's Assistant*,
   concludes its problems on double false position with the following:

   When first the marriage knot was ty'd
   Between my wife and me,

   My age was to that of my bride
   As three times three to three

   But now when ten and half ten years,
   We man and wife have been,

   Her age to mine exactly bears,
   As eight is to sixteen;

   Now tell, I pray, from what I've said,
   What were our ages when we wed?

   Answer: Thy age when marry'd must had been
                    Just forty-five; thy wife's fifteen.

   Check the answer. Derive it by false position and by simple algebra.

1-11 GEOMETRY

We can summarize Egyptian geometry quite briefly. That which has
been found consists of a number of problems in which the inclination
of a line and a plane was determined, or volumes and areas of mathematical figures were computed. The solutions of these problems proceeded according to definite arithmetical instructions, some of which are
correct and some of which are not. The calculations of the area of a
rectangle, a triangle, and a trapezoid were all correct. The area of an
arbitrary quadrilateral was calculated as the product of half the sum of
two opposite sides and half the sum of the other two, and therefore
the procedure was incorrect. There is no record of statements of general
theorems or of proofs; the chief concern of the Egyptians seemed to be
to obtain a useful result.

DABOLL'S
SCHOOLMASTER'S ASSISTANT.

IMPROVED AND ENLARGED.

BEING A

PLAIN PRACTICAL SYSTEM

OF

ARITHMETICK.

ADAPTED TO

THE UNITED STATES.

BY NATHAN DABOLL.

WITH THE ADDITION OF THE

FARMERS' AND MECHANICKS' BEST

METHOD OF BOOK-KEEPING,

DESIGNED AS A

COMPANION TO DABOLL'S ARITHMETICK.

BY SAMUEL GREEN.

ITHACA, N. Y.,

PRINTED AND PUBLISHED BY MACK, ANDRUS, AND WOODRUFF

1837.

5. A Goldsmith sold 1 lb. of gold, at 2 cts. for the first ounce, 8 cents for the second, 32 cents for the third, &c. in a quadruple proportion geometrically: what did the whole come to? *Ans. $1118·18, 10 cts.*

6. What debt can be discharged in a year, by paying 1 farthing the first month, 10 farthings, or (2½d) the second and so on, each month in a tenfold proportion? *Ans. £115740740 14s. 9d. 3 qrs.*

7. A thrasher worked 20 days for a farmer, and received for the first days work four barley-corns, for the second 12 barley corns, for the third 36 larley corns, and so on, in triple proportion geometrically. I demand what the 20 day's labour came to supposing a pint of barley to contain 7680 corns, and the whole quantity to be sold at 2s. 6d. per bushel? *Ans. £1773 7s. 6d. rejecting remainders*

8. A man bought a horse, and by agreement, was to give a farthing for the first nail, two for the second, four for the third, &c. There were four shoes, and eight nails in each shoe; what did the horse come to at that rate? *Ans. £4473924 5s. 3¾d*

9. Suppose a certain body, put in motion, should move the length of 1 barley-corn the first second of time, one inch the second, and three inches the third second of time, and so continue to increase its motion in triple proportion geometrical; how many yards would the said body move in the term of half a minute. *Ans. 95319965623 yds. 1ft. 1 in. 1b. which is no less than five hundred and forty-one millions of miles.*

POSITION.

POSITION is a rule which, by false or supposed numbers, taken at pleasure, discovers the true ones required.—It is divided into two parts, Single or Double.

SINGLE POSITION

IS when one number is required, the properties of which are given in the question.

RULE.—1. Take any number and perform the same operation with it, as is described to be performed in the question.

2. Then say; as the result of the operation : is to the given num in the question : : so is the supposed number : to the true one required.

The method of proof is by substituting the answer in the question.

EXAMPLES.

1. A schoolmaster being asked how many scholars he had, said, If I had as many more as I now have, half as many, one-third, and one fourth as many, I should then have 148; How many scholars had he?

Suppose he had 12   As 37 : 148 : : 12 : 48 *Ans.*

as many = 12		48
½ as many = 6		24
⅓ as many = 4		16
¼ as many = 3		12
Result, 37		Proof, 148

2. What number is that which being increased by ⅓, ¼, and ⅕ of itself, the sum will be 125? *Ans. 60.*

3. Divide 93 dollars between A, B and C, so that B's share may be half as much as A's, and C's share three times as much as B's. *Ans. A's share $31, B's $15½, and C's $46½.*

4. A, B and C, joined their stock and gained 360 dols. of which A took up a certain sum, B took 3½ times as much as A, and C took up as much as A and B both; what share of the gain had each? *Ans. A $40, B $140, and C $180.*

5. Delivered to a banker a certain sum of money, to receive interest for the same at 6d. per cent. per annum, simple interest, and at the end of twelve years received 731l, principal and interest together; what was the sum delivered to him at first? *Ans. £425.*

6. A vessel has 3 cocks, A, B and C; A can fill it in 1 hour, B in 2 hours, and C in 4 hours; in what time will they all fill it together? *Ans 34 min. 17¼ sec.*

(d)

DOUBLE POSITION,

TEACHES to resolve questions by making two suppo
sitions of false numbers.*

RULE.

1. Take any two convenient numbers, and proceed with
each according to the conditions of the question.

2. Find how much the results are different from the results in the question.

3. Multiply the first position by the last error, and the last
position by the first error.

4. If the errors are alike, divide the difference of the products by the difference of the errors, and the quotient will
be the answer.

5. If the errors are unlike, divide the sum of the products by the sum of the errors, and the quotient will be
the answer.

NOTE.—The errors are said to be alike when they are
both too great, or both too small; and unlike, when one
is too great, and the other too small.

EXAMPLES.

1. A purse of 100 dollars is to be divided among 4 men
A, B, C and D, so that B may have four dollars more than
A, and C 8 dollars more than B, and D twice as many as
C; what is each one's share of the money?

```
1st. Suppose A   6          2d. Suppose A    8
             B  10                      B   12
             C  18                      C   20
             D  36                      D   40
                ---                        ---
                 70                         80
                100                        100
                ---                        ---
   1st error,    30         2d error,       20
```

* Those questions in which the results are not proportional to their positions, belong to this rule; such as those in which the number sought is increased or diminished by some given number, which is no known part of the
number required.

(e)

The errors being alike, are born too small, therefore,

```
Pos.  Err.
 6    30
 8    20
       X
240   120
240   120
```

```
A  $12
B   16
C   24
D   48
Proof 100
```

10)120(12 A's part.

2. A, B, and C, built a house which cost 500 dollars, of
which A paid a certain sum; B paid 10 dollars more than
A and C paid as much as A and B both; how much did
each man pay?
Ans. A paid $120, B $130, and C $250.

3. A man bequeathed 100l. to three of his friends, after
this manner; the first must have a certain portion, the second must have twice as much as the first, wanting 8l. and
the third must have three times as much as the first, wanting 15l.; I demand how much each man must have?
Ans. The first £20 10s. second £33, third £46 10s.

4. A labourer was hired for 60 days upon this condition;
that for every day he wrought he should receive 4s. and for
every day he was idle should forfeit 2s.; at the expiration
of the time he received 7l. 10s.; how many days did he
work, and how many was he idle?
Ans. He wrought 45 days, and was idle 15 days.

5. What number is that which being increased by its ¼,
its ⅓, and 18 more, will be doubled?   Ans. 72.

6. A man gave to his three sons all his estate in money,
viz. to F half, wanting 50l., to G one-third, and to H the
rest, which was 10l. less than the share of G; I demand
the sum given, and each man's part?
Ans. the sum given was £360, whereof F had £130 G
£120, and H £110

Figure 1-5 Title page and the Rules of
False Position from *Daboll's Schoolmas-
ter's Assistant.* This was the most popular
arithmetic in America prior to 1850. This
English textbook was first reprinted in
New London, Connecticut, in 1800 and
later revised for American use by David A.
Daboll.

Their process for finding the area of a circle with diameter 9 is illustrated by Problem 50 of the Rhind papyrus:

Example of a round field of a diameter 9 khet. What is its area? Take away $\frac{1}{9}$ of the diameter, 1; the remainder is 8. Multiply 8 times 8; it makes 64. Therefore it contains 64 setat of land.

We see that the method amounts to the use of the formula

$$A = (d - \frac{1}{9}d)^2,$$

with $d = 9$. The derivation of this expression may be imagined as follows (see Figure 1-6). Let the square have a side of length d. Then its

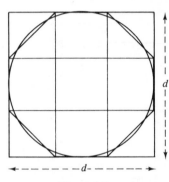

area is d^2. The square can be divided into nine smaller squares, as shown in the figure. The area of each of these squares is $\frac{1}{9}d^2$. The area of the circle is approximately equal to the area of seven small squares, that is, $7 \cdot \frac{1}{9}d^2$, which is equal to $\frac{63}{81}d^2$. If we take $\frac{64}{81}d^2$ instead of $\frac{63}{81}d^2$, we are not so far off the mark, and $\frac{64}{81}d^2$ has the practical advantage of being a perfect square, $(\frac{8}{9}d)^2$. This can be written as $(d - \frac{1}{9}d)^2$, the desired result.

Figure 1-6

The modern formula for the area of a circle is πr^2, in which r is the length of the radius. In Figure 1-8, $r = \frac{1}{2}d$, and hence

$$\text{area circle} = \pi (\frac{1}{2}d)^2 = \frac{1}{4}\pi d^2.$$

Now if $\frac{1}{4}\pi d^2 = \frac{64}{81}d^2$, then $\frac{1}{4}\pi = \frac{64}{81}$. Therefore, the Egyptian algorithm is equivalent to the approximation

$$\pi = \frac{256}{81} = 3.16 \ldots,$$

which was not a bad value since it was not far from the actual value of 3.14.... However, we must keep in mind that the Egyptian method did not use the idea of a constant such as π.

The *pyramids* are often considered visual proof of the Egyptian proficiency in mathematics. Rather than assume that all our readers are skilled in solid geometry, we shall describe an Egyptian pyramid. Suppose that, in Figure 1-7, the quadrilateral $ABCD$ is a square situated in a horizontal plane, in this case the ground, and that $\overline{TT'}$ is the perpendicular erected at the point of intersection of the diagonals of the square. Point T is joined to A, B, C, and D. The resulting figure $TABCD$ is a square pyramid. The square $ABCD$ is called the *base*; $\overline{TA}$, $\overline{TB}$, $\overline{TC}$, $\overline{TD}$, and the sides of the square are the *edges*. TAB, TBC, TCD, and TDA are the *lateral faces*. The line segment $\overline{TT'}$ is called the *altitude* of the pyramid.

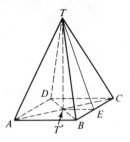

Figure 1-7

The Egyptians knew that the volume of a pyramid is equal to $\frac{1}{3}B \times a$, where B is the area of the base and a the length of the altitude. Therefore, if they knew the side of the base and also the altitude, they could compute the volume. They also calculated a number, the *skd* (pronounced "sayket"), which appears to be the ratio of half the side of the base to the altitude, $T'E/TT'$. Today, we would call this the cotangent of the angle of inclination, TET', of a face of the pyramid.

In the Moscow papyrus we find the equivalent of a formula representing the volume of the solid shown in Figure 1-8. The quadrilaterals $ABCD$ and $EFGH$ are squares, and $\overline{CG}$ is perpendicular to the parallel planes $ABCD$ and $EFGH$. The lines $\overleftrightarrow{AE}$, $\overleftrightarrow{BF}$, $\overleftrightarrow{CG}$, and $\overleftrightarrow{DH}$ pass through one point. This solid is called a *truncated pyramid*. Its volume is given as

$$\frac{1}{3}h\,(a^2 + b^2 + ab),$$

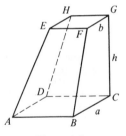

Figure 1-8

in which a, b, and h represent the lengths BC, FG, and CG, respectively. The formula is perfectly correct but is given without a derivation. It is difficult to assume that such a calculation of a volume was not reasoned out in some way or other. However, until further information about Egyptian methods is found, an explanation of their derivation is mere guesswork. Figure 1-9 shows this problem.

Egyptian mathematics, then, has often been described as largely motivated by practical needs and developed by empirical processes. However, some rather remarkable geometric results were obtained. In arithmetic, the repeated solution of *aha* problems by the same process testifies to the existence of some generalizations, even though they were never stated explicitly or proved. Further, several of the problems in the Rhind papyrus, such as Exercise 2 of Exercises 1-10, suggest that the Egyptians also developed interests of a theoretical or recreational nature.

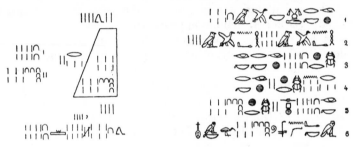

Figure 1-9 Calculation of the volume of a truncated pyramid. (From W. W. Struve, "Mathematischer Papyrus des Staatlichen Museums der Schönen Künste in Moskau," *Quellen und Studien zur Geschichte der Mathematik*, Part A, *Quellen*, 1 (1930).) The upper half of the picture is a photographic reproduction of a part of the Moscow papyrus. It was written in hieratic. In the lower half the reproduction is translated into hieroglyphics. From the left half of this translation we can easily follow the calculation of the volume. The drawing shows that $a = 4$, $b = 2$, and $h = 6$ (see page 37). According to the formula, the volume is

$$\frac{1}{3} \cdot 6 \cdot (4^2 + 2^2 + 4 \cdot 2) = 56.$$

In the text (which should be read from right to left) one actually finds:

$$2^2 = 4, \qquad \frac{1}{3} \cdot 6 = 2, \qquad 4^2 = 16.$$

To this 16 are added 8 and 4, which gives 28 (lower left; in the hiero-glyphic translation next to the "papyrus roll"). In the upper left this 28 is then multiplied by 2, producing the final result, 56.

Problem 79 of the Rhind papyrus is a second example of this kind and is the only known example of a geometric progression in Egyptian mathematics. It reads: "Sum the geometric progression of five terms, of which the first term is 7 and the multiplier is 7." The problem is solved in two ways in the papyrus:

1. Solution by multiplying			*2. Solution by adding*	
7 ✕	2,801		houses	7
\ 1	2,801		cats	49
\ 2	5,602		mice	343
\ 4	11,204		spelt	2,401
			hekat	16,807
Total	19,607		Total	19,607

The first column suggests that the Egyptians had the equivalent of a recursive relation for the sum of a geometric progression in which the first term and the common ratio are the same. If r represents the first term as well as the ratio, and S_n represents the sum of the first n terms of a geometric progression, we have

$$\begin{aligned}
S_n &= r + r^2 + r^3 + \cdots + r^{n-1} + r^n \\
&= r(1 + r + r^2 + \cdots + r^{n-2} + r^{n-1}) \\
&= r(1 + (r + r^2 + \cdots + r^{n-1})) \\
&= r(1 + S_{n-1}) \\
&= r(S_{n-1} + 1).
\end{aligned}$$

If we now assign the values $1, 2, \ldots, 5$ to n, we get

$$\begin{aligned}
S_1 &= 7(S_0 + 1) = 7(0 + 1) = 7 \\
S_2 &= 7(S_1 + 1) = 7 \cdot 8 = 56 \\
S_3 &= 7(S_2 + 1) = 7 \cdot 57 = 399 \\
S_4 &= 7(S_3 + 1) = 7 \cdot 400 = 2800 \\
S_5 &= 7(S_4 + 1) = 7 \cdot 2801 = 19,607.
\end{aligned}$$

Column (1) in the papyrus suggests that the writer used a similar method and gave only the last step of the calculation. This last step is the calculation of 7×2801. The second column merely gives the straightforward sum of all the terms. The words "houses," "cats," and so on, have never been fully explained. One theory is that they were merely names for the powers of 7. Another is that this problem was a puzzle problem, similar to several familiar nursery stories and rhymes. Perhaps the idea was, that if there were seven houses, each with seven cats, each of which killed seven mice, each of which would have eaten seven ears, each of which would have produced seven hekat of grain, how much grain was saved? A similar problem is found in Leonardo of Pisa's *Liber Abaci* (1202), and some people have been reminded of the rhyme

> As I was going to St. Ives,
> I met a man with seven wives,
> Every wife had seven sacks,
> Every sack had seven cats,
> Every cat had seven kits.
> Kits, cats, sacks, and wives,
> How many were there going to St. Ives?

EXERCISES 1-11

1   Find the sum of the first five terms of the geometric progression with first term and common ratio both equal to 6. Use
    (a) the method of recursion (see page 39),
    (b) the method of addition (see page 39).

2   Repeat Exercise 1 with the first term and common ratio equal to $\frac{1}{2}$ instead of 6.

3   Show that the ancient Egyptian procedure for finding the area of a quadrilateral gives a correct result if the quadrilateral is a rectangle and gives too large a number if the figure is a nonrectangular parallelogram or a trapezoid. Is the procedure ever correct for a quadrilateral that is not a rectangle?

4   (a) Use the ancient Egyptian procedure to find the area of a circle with diameter 12.
    (b) What is the area of the circle in (a) if we assume that $\pi = 3.14$?
    (c) Assuming that the answer in (b) is correct, find the percent error of the answer in (a).

5   R. J. Gillings, on pages 139-146 of reference 2 (this chapter), compares several explanations of the ancient Egyptian method for calculating the area of a circle. He believes that Problem 48 of the Rhind papyrus should be regarded as a general procedure containing the concept of a proof. Read Gillings' discussion and use similar diagrams and reasoning to justify the Egyptian value of the area of the circle that you found in Exercise 4.

6   Check the translation of the hieroglyphic numerals and computation in the lower-left portion of Figure 1-9.

REFERENCES

[1]   Chace, A. B., and others, *The Rhind Mathematical Papyrus*. Oberlin, Ohio: Mathematical Association of America, 1927-1929, 2 vols. Contains an annotated translation of the Rhind papyrus.

[2]   Gillings, Richard J., *Mathematics in the Time of the Pharaohs*. Cambridge, Mass.: The MIT Press, 1972.

[3]   Midonick, Henrietta D., *The Treasury of Mathematics*. New York: Philo-
      sophical Library, Inc., 1965; Baltimore: Penguin Books, Inc., 1968, 2 vols.
      (paperback). Contains substantial extracts from the Moscow and Rhind
      papyri.

[4]   Neugebauer, Otto, *The Exact Sciences in Antiquity*, 2nd ed. Providence, R. I.:
      Brown University Press, 1957; New York: Harper & Row, Publishers, 1962
      (paperback).

[5]   Neugebauer, Otto, *Geschichte der antiken mathematischen Wissenschaften.
      Erster Band, Vorgriechische Mathematik*. New York: Springer-Verlag New
      York, 1969.

[6]   Newman, James R., "The Rhind Papyrus," *The World of Mathematics*,
      New York: Simon and Schuster, 1956, 4 vols.: Vol. 1, pp. 170-178. Contains
      a discussion of a few problems from the Rhind papyrus.

[7]   Van der Waerden, B. L., *Science Awakening*, tr. by Arnold Dresden.
      Groningen, The Netherlands: Wolters-Noordhoff, 1954; New York:
      John Wiley & Sons, Inc., 1963 (paperback).

      For further data on continued fractions, see

[8]   Moore, Charles G., *An Introduction to Continued Fractions*. Washington,
      D. C.: National Council of Teachers of Mathematics, 1964.

[9]   Olds, C. D., *Continued Fractions*. New York: Random House, Inc., 1963
      (Volume 9 of The New Mathematical Library).

      More information with reference to mathematics in prehistoric times and
      mathematical concepts observed in nature can be found in Volume 1 of *The World
      of Mathematics* (see reference 6), which contains essays on counting and "count-
      ing" by birds, and in references 10 through 14.

[10] Seidenberg, A., *The Diffusion of Counting Practices*. Berkeley, Calif.: Univer-
      sity of California Press, 1960.

[11] Seidenberg, A., "The Ritual Origin of Geometry," *Archive for History of
      Exact Sciences*, Vol. I (1960-1962), pp. 488-527.

[12] Seidenberg, A., "The Ritual Origin of Counting," *Archive for History of
      Exact Sciences*, Vol. 2 (1962-1966), pp. 1-40.

[13] Struik, D. J., *A Concise History of Mathematics*. New York: Dover Publica-
      tions, Inc., 1948, 2 vols.

[14] Vogel, Kurt, *Vorgriechische Mathematik*, Teil I. Hannover, West Germany:
      Hermann Schroedel Verlag KG, 1958.

      For further historical data on unit fractions, see

[15] Dickson, Leonard E., *History of the Theory of Numbers*. New York: G. E.
      Stechert, 1934, 3 vols.

2

BABYLONIAN MATHEMATICS

2-1 SOME HISTORICAL FACTS

It was not a mere chance that the Egyptian civilization grew up in the valley of the Nile. The river furnished both transportation and the water needed to make the land productive. Two rivers, the Tigris and the Euphrates, supported the Babylonian civilization. Look at a map to see why historians have called the lands of the early Hebrew, Phoenician, and Babylonian civilizations "the fertile crescent."

Between 3000 and 2000 years B.C. the southern part of Mesopotamia was ruled by the Sumerians, whose culture had reached a high level. They belonged to the first peoples who were able to write. They wrote by pressing their symbols into the surface of soft clay tablets with a wedge-shaped stylus. The tablets were then baked. Many of them have been unearthed within the last century. The wedge-shaped (*cuneiform*) symbols were also used to represent numerals.

The oldest texts date from about 3000 B.C., the time of the first dynasty of Ur. In the course of time a people living more to the north, the Akkadians, migrated to the south. They eventually dominated the Sumerians and took over much of their higher culture, including their numeration system.

About 1800 B.C., Hammurabi, the king of the city of Babel, came into power over the whole empire of Sumer and Akkad and founded the first Babylonian dynasty. After his death, his empire disintegrated, but its culture remained for many years.

The oldest Babylonian mathematical texts known to us date from the period 1900-1600 B.C.

2-2 BABYLONIAN NUMERICAL NOTATION

To gain an insight into Babylonian arithmetic, it is necessary to understand positional numeration systems. For this purpose, we look again at the ancient Egyptian notation. If we were to write

$$\text{II} \, \text{ℰℰℰ} \, {}^{\text{∩∩∩∩}}_{\text{∩∩∩}} \quad \text{instead of} \quad \text{ℰℰℰ} \, {}^{\text{∩∩∩∩}}_{\text{∩∩∩}} \, \text{II},$$

the number represented would remain 372. But if in our modern notation we were to write 237 in place of 372, we would have written something quite different.

From this example we see that in our system the number represented by a *digit* (one of the symbols 0, 1, 2, 3, 4, 5, 6, 7, 8, 9) in a numeral is determined by the *place* (or *position*) of that digit in the numeral. In 237, "3" represents 3 times 10, and in 372 it represents 3 times 10^2.

In Egyptian notation, if more than one 10 or more than one 100 was to be represented, the symbol ∩ or ℰ was repeated. That is, the Egyptians used a *repetitive principle*. However, in this system repeated symbols were collected into groups of 10 and these groups replaced by single symbols of the next larger denomination. Thus, 10 was the *base* of the Egyptian numeration system. In evaluating Egyptian numerals, we simply add the numbers represented by the symbols. In other words, the Egyptians also used an *additive principle*.

In evaluating a numeral in our system of numeration, 243 for example, we proceed as follows:

$$243 = 2 \cdot 10^2 + 4 \cdot 10 + 3.$$

Our system does not use the repetitive principle of the Egyptians. Rather, the 10 is multiplied by 4, the digit in the second place from the right, instead of being written four times. Similarly, 10^2 is multiplied by 2 instead of being written twice. Thus, our system uses a *multiplicative principle*. We also note that the power of the base (in this case 10) by which each digit is multiplied is determined solely by the position of the digit in the numeral. Thus, our system of numeration is a *positional system*.

Both our system and the Egyptian system use 10 as a base, but our system has a different symbol for each of the numbers 0 through 9.

If the numeral 243 were given in base a, we could write the following:

$$243_a = 2 \cdot a^2 + 4 \cdot a + 3.$$

In decimal numeration we need 10 digits, one for each whole number less than 10. If we take 7 as a base, we need only the digits 0, 1, 2, 3, 4, 5, and 6. The number 7 and larger numbers can be represented by combinations of these digits, for we have:

$7 = 1 \cdot 7 + 0$; hence in the base 7 system, 7 becomes 10_7 (pronounced "one-zero base seven");

$8 = 1 \cdot 7 + 1$; hence in the base 7 system, 8 becomes 11_7 (pronounced "one-one base seven");

$9 = 1 \cdot 7 + 2$; hence in the base 7 system, 9 becomes 12_7 (pronounced "one-two base seven").

We shall explore these other bases, their history, and their modern uses in Chapter 8.

After this introduction we are prepared to understand the Babylonian system of numeration. The Babylonians used an incomplete *sexagesimal positional system.* "Sexagesimal" means "with base 60." A complete sexagesimal positional system would need a symbol for zero and for 59 other digits. However, the Babylonians had no symbol for zero, and the other 59 digits were written as combinations of only two different marks:

the wedge Υ for the units

the corner $\langle$ for the tens.

With these two marks they wrote numbers less than 60 as follows:

1	2	3	4	5	6	7	8	9

10	11	12	20	30	40	50	59

These first 59 symbols were used as "digits" in a sexagesimal positional system. When 60 was reached, these 59 symbols were used again to denote the number of sixties. Thus,

1 sixty was represented by one wedge, Υ

2 sixties by two wedges, $\Upsilon\!\Upsilon$

10 sixties by one corner,

59 sixties by five corners and nine wedges, ⟨⟨⟨⟨⟨𝖄.

We observe that in writing the first 59 numbers, the Babylonians followed the same method as the Egyptians, with the difference that they used a wedge for the unit and a corner for the ten. In both the ones position and the sixties position, the numbers are represented in base 10.

The same procedure is used in the 60^2's position, the 60^3's position, and so on. Thus,

represents

$$11 \cdot 60^2 \ + \ 23 \cdot 60 \ + \ 32 \qquad (= 41{,}012).$$

In Babylonian numerical notation the units symbols are written on the right, to the left of them the symbols for the number of sixties, to the left of those the symbols for the number of sixty squared's, and so on. Modern scholars use Hindu-Arabic numerals, set off by commas, to represent Babylonian symbols for the whole numbers. They use a semicolon to separate the symbols for fractions from those for whole numbers. We shall call such a representation a sexagesimal numeral. Consider the examples in Table 2-1.

Table 2-1

	Decimal	Sexagesimal	Babylonian
	63	1,3	
	132	2,12	
	1547	25,47	
	$2\frac{1}{2} = 2\frac{30}{60}$	2;30	
	$\frac{3}{4} = \frac{45}{60}$	0;45	

There are three sources of ambiguity in the Babylonian numeration system. One is the fact that the Babylonians did not have a "sexagesimal point." That is, they had no mark corresponding to the semicolon that we have used in writing sexagesimal fractions and mixed numbers. Thus, one cannot be certain if ▌▌ represents 2, or $1 + \frac{1}{60}$, or $\frac{2}{60}$, or, in fact, still other numbers. The examples in Table 2-2 illustrate two other types of ambiguities:

Table 2-2

	Decimal	Sexagesimal	Babylonian
	12	12	⟨𝕀𝕀
	602	10,2	⟨𝕀𝕀
	1	1	𝕀
	60	1,0	𝕀
	7236	2,0,36	𝕀𝕀 ⟪ 𝖄𝖄𝖄
	156	2,36	𝕀𝕀 ⟪ 𝖄𝖄𝖄

In the table there is no difference between the Babylonian notation in the first line and the second line, because there are no noncomposite symbols for the digits 1 through 59 (except for 1 and 10). The meaning of ⟨ 𝕀 𝕀 is uncertain: Is it $10 + 2 = 12$, or $10 \cdot 60 + 2 = 602$, or $11 \cdot 60 + 1 = 661$, or something else?

There is no difference between the Babylonian symbols in the third and fourth lines above because there is no zero. For a similar reason there is an ambiguity in the Babylonian symbols in the fifth and sixth lines. In a later Babylonian period (the last few centuries B.C.) a symbol ⟨ was used for the zero that occurred between two digits. But there was never a symbol for a terminal zero. Therefore, in cases where a terminal zero would have been useful, the number represented had to be determined by the rest of the calculation or by the context of the problem.

Additional examples of fractions are given in Table 2-3.

Table 2-3

Decimal	Sexagesimal	Babylonian
$\frac{1}{5} = \frac{12}{60}$	0;12	⟨𝕀𝕀
$\frac{2}{27} = \frac{4}{60} + \frac{26}{60^2} + \frac{40}{60^3}$	0;4,26,40	𝖄𝖄 ⟪𝖄𝖄𝖄 ⟨⟨⟨
$1\frac{3}{8} = 1 + \frac{22}{60} + \frac{30}{60^2}$	1;22,30	𝕀 ⟪𝕀𝕀 ⟪⟪

The reader has often met Babylonian ideas, perhaps without realizing it. Biblical references to the Babylonians are frequent. The following short Babylonian table of weights and money,

$$1 \text{ talent} = 60 \text{ mina}$$
$$1 \text{ mina} = 60 \text{ shekel,}$$

contains familiar words defined in a sexagesimal context.

The words "minute" and "second" also stem, indirectly, from the Babylonians. In condensed form, the story is as follows. The Babylonians developed an interest in astronomy in connection with constructing a calendar to help with the yearly cycle of planting and harvesting. Greek astronomers used Babylonian data, acquired through the contacts of trade and conquest. These data were expressed in the sexagesimal numeration system. The Greeks adopted this system for writing the fractions they used in astronomy, calling sixtieths the "first small parts," sixtieths of sixtieths the "second small parts," and so on. This terminology was retained when the Greek astronomical treatises were translated into Arabic and when the Arabic manuscripts were later (in the twelfth century) translated into Latin by Western European scholars. Sixtieths and sixtieths of sixtieths became *pars minuta prima* and *pars minuta secunda*, respectively. When these phrases were shortened as they were translated into English, the words *minutes* and *seconds*, used in measuring angles and time, resulted.

Figure 2-1 is a photograph of a famous cuneiform tablet containing only numbers. They are easily recognized.

We summarize the Babylonian system of numeration as follows:

1. The numbers were written in a sexagesimal positional system.
2. The digits of this system, the numerals 1 through 59, were constructed additively from the marks Υ and $\langle$ in base 10.
3. There was no sexagesimal point and no zero.

EXERCISES 2-2

1 Give three numbers that might be represented by

   (a) $\Upsilon\Upsilon$ (b) $\langle\Upsilon\Upsilon$

2 Assume that the symbol $\Upsilon \langle\langle\Upsilon \langle\langle\langle$ represents 1;22,30.

   (a) Find the numeral for the number in decimal notation.

   (b) Give at least one other number that the Babylonian symbol could represent.

3 (a) Multiply the sexagesimal number 1;22,30 by 60.

   (b) Describe a general procedure for multiplying a sexagesimal number by 60.

Figure 2-1 Babylonian cuneiform tablet containing tables with numbers (catalog number 322 from the G. A. Plimpton Collection, Columbia University). From O. Neugebauer and A. Sachs, *Mathematical Cuneiform Texts*, American Oriental Society, 1946.)

4 Write in cuneiform:
  (a) 29   (b) 43
  (c) 78   (d) 100
  (e) 577   (f) 4405

5 Write in cuneiform:
  (a) $\frac{1}{2}$ (b) $\frac{3}{4}$

  (c) $\frac{5}{6}$ (d) $\frac{1}{8}$

  (e) $1\frac{4}{9}$ (f) $86\frac{1}{90}$

(Hint: See Section 8-9.)

2-3 THE FUNDAMENTAL OPERATIONS

Addition and subtraction using Babylonian numerals presented no difficulties. The computation proceeded as in our decimal system, but instead of borrowing or carrying tens the Babylonian borrowed or carried sixties. Multiplication was performed as in our modern system, but the Babylonians needed multiplication tables in which the products 1×2, $2 \times 2, \ldots$, 59×2, 1×3, $2 \times 3, \ldots$, $59 \times 3, \ldots$, 1×59, $2 \times 59, \ldots$, 59×59 occurred. Several of these tables have been discovered on clay tablets. (In reality, the tables were shorter. Thus, the "2's table" consisted only of 1×2, $2 \times 2, \ldots$, 19×2, 20×2, 30×2, 40×2, 50×2. To find 26×2, for instance, one only had to look up 20×2 and 6×2 in this table and add the results.)

Division was accomplished by multiplying the dividend by the multiplicative inverse, or reciprocal, of the divisor. To calculate $47 \div 3$, for instance, the Babylonians first calculated $1 \div 3$ and then multiplied the result by 47.

To simplify these calculations, tables of reciprocals were constructed. One of the oldest known tables contains the reciprocals of those counting numbers which are products of factors of 60, up to and including the number 81. These are the reciprocals that can be written as terminating sexagesimal fractions. The beginning of the table reads:

$$
\begin{array}{ll}
1 \div 2 = 0;30 & 1 \div 6 \; = 0;10 \\
1 \div 3 = 0;20 & 1 \div 8 \; = 0;7,30 \\
1 \div 4 = 0;15 & 1 \div 9 \; = 0;6,40 \\
1 \div 5 = 0;12 & 1 \div 10 = 0;6.
\end{array}
$$

Now, how did the Babylonians write the reciprocals of the other numbers, such as $1 \div 7$, $1 \div 11$? As a rule they avoided the use of these fractions. Occasionally, they approximated such fractions (as we do in approximating $1 \div 7$ by 0.14 in modern notation.)

In addition to tables of reciprocals, there were also tables listing multiples of reciprocals. An example follows.

Multiplication table for   0;6,40 (= 1 ÷ 9):

$$1 \times 0;6,40 = 0;6,40 \qquad 8 \times 0;6,40 = 0;53,20$$
$$2 \times 0;6,40 = 0;13,20 \qquad \vdots$$
$$3 \times 0;6,40 = 0;20 \qquad 19 \times 0;6,40 = 2;6,40$$
$$4 \times 0;6,40 = 0;26,40 \qquad 20 \times 0;6,40 = 2;13,20$$
$$5 \times 0;6,40 = 0;33,20 \qquad 30 \times 0;6,40 = 3;20$$
$$6 \times 0;6,40 = 0;40 \qquad 40 \times 0;6,40 = 4;26,40$$
$$7 \times 0;6,40 = 0;46,40 \qquad 50 \times 0;6,40 = 5;33,20.$$

EXAMPLE 1. $5 \div 9 = 5 \times (1 \div 9) = 5 \times 0;6,40 = 0;33,20.$

EXERCISES 2-3

1   Complete the table of reciprocals up to $1 \div 60$. (List only those reciprocals that can be written as terminating sexagesimal fractions.)

2   Complete the table of 0;6,40 for multipliers 9 through 18, and then calculate:
   (a) $7 \div 9$ (b) $13 \div 9$
   (c) $25 \div 9$ (d) $47 \div 9$

3   Calculate in the Babylonian way:
   (a) $18 \div 5$ (b) $23 \div 12$
   (c) $17 \div 10$ (d) $9 \div 8$
   (e) $5 \div 16$ (f) $7 \div 24$

2-4   EXTRACTION OF ROOTS

To find the square root of a number, the Babylonians used a method of approximation. Their probable line of thought can be described by means of a few examples.

EXAMPLE 1 $\sqrt{17}$ is a little more than 4. Hence, 4 is too small.

Now we have $\sqrt{17} \cdot \sqrt{17} = 17$

and also $4 \cdot \dfrac{17}{4} = 17.$

Since, in the last product, the first factor,

$$4, \text{ is less than} \sqrt{17},$$

the second factor,

$$\frac{17}{4}, \text{ is greater than } \sqrt{17}.$$

Now, as an approximation to $\sqrt{17}$, the Babylonians chose the average of the two numbers 4 and $\frac{17}{4}$. Therefore, they obtained

$$\sqrt{17} \approx \frac{1}{2}(4 + 4\frac{1}{4}) = 4\frac{1}{8}.$$

EXAMPLE 2   In a Babylonian text we find: $\sqrt{2} = 1\frac{5}{12}$. This result can be explained in the same way. As a first approximation to $\sqrt{2}$, we choose a number such that its square is close to 2. Such a number is 1. Now $2 \div 1 = 2$. The average of 1 and 2 is $1\frac{1}{2}$, the second approximation. Since $(1\frac{1}{2})^2 = 2\frac{1}{4}$, which is greater than 2, the value $1\frac{1}{2}$ is too large. Therefore, $\frac{2}{1\frac{1}{2}}$, which is equal to $1\frac{1}{3}$, is too small. As a third approximation to $\sqrt{2}$, we choose the average of $1\frac{1}{2}$ and $1\frac{1}{3}$. Hence,

$$\sqrt{2} \approx \frac{1}{2}(1\frac{1}{2} + 1\frac{1}{3}) = 1\frac{5}{12}.$$

EXERCISES 2-4

1 See Example 2. Find in the same way a first, a second, and a third approximation to
   (a) $\sqrt{10}$ (b) $\sqrt{7}$
   (c) $\sqrt{15}$ (d) $\sqrt{27}$

2 The following result occurs in a Babylonian text:

$$\sqrt{40^2 + 10^2} = 40 + \frac{100}{2 \cdot 40}.$$

Try to get this result yourself. (Hint: Start by observing that $\sqrt{40^2 + 10^2}$ must be slightly greater than 40.)

2-5 BABYLONIAN ALGEBRA

The Babylonians solved a *system of linear equations* with two or more unknowns just as we do.

EXAMPLE 1   Of two unknowns, the length l and width w of a rectangle, it is given that

$$\begin{cases} l + \dfrac{1}{4}w = 7 \\ l + w = 10. \end{cases}$$

Calculate the length and the width.

Solution

| *Babylonian* | *Modern* |
| | |

$7 \times 4 = 28$

$$\begin{cases} 4l + w = 28 \\ l + w = 10 \end{cases}$$

$28 - 10 = 18$ $3l = 18$

$18 \times \dfrac{1}{3} = 6$ (the length) $l = 6$

$10 - 6 = 4$ (the width). $w = 10 - 6 = 4.$

This example shows that the Babylonians followed the same sequence of steps that we do, but they did not use letter symbols to represent numbers. We have no knowledge of their reasoning processes, only of the calculations that resulted.

The Babylonians also solved problems of the following type: *Find the numbers the sum (or the difference) of which and the product of which are given.*

EXAMPLE 2   Find two numbers that have a sum of 14 and a product of 45.

We give the solution in modern notation. Two numbers the sum of which is 14 can always be written in the form $7 + a$ and $7 - a$. Then we have

$$(7 + a)(7 - a) = 45$$
$$49 - a^2 = 45$$
$$a^2 = 4$$
$$a = 2.$$

Hence, the numbers are 9 and 5.

Using the process just discussed, the Babylonians were also able to solve *quadratic equations*.

EXAMPLE 3   Solve: $x^2 + 6x = 16$.

Solution   The Babylonian solution amounts to the following, in modern notation. Write the equation in the form

$$x(x + 6) = 16.$$

Let $x + 6 = y.$

Then we must find two numbers x and y such that

$$y - x = 6 \quad \text{and} \quad xy = 16.$$

Now let $y = a + 3$ and $x = a - 3.$

Then

$$(a + 3)(a - 3) = 16$$
$$a^2 - 9 = 16$$
$$a = 5.$$

Hence, $x = 2.$

In the quadratic equation of Example 3, the coefficient of x^2 is equal to 1. The Babylonians were also able to solve quadratic equations in which the coefficient of x^2 is not 1.

EXAMPLE 4   Solve: $7x^2 + 6x = 1.$ (1)

We could reduce this equation to the foregoing type by dividing both members by 7. We would then get

$$x^2 + \frac{6}{7}x = \frac{1}{7}.$$

The Babylonians did not do this. They preferred to avoid such fractions as $\frac{6}{7}$ and $\frac{1}{7}$, which could not be represented by a terminating sexagesimal fraction. For this reason, they multiplied both terms of equation (1) by 7. This results in

$$(7x)^2 + 6(7x) = 7.$$

They now treated $7x$ as if it were the unknown. If we represent $7x$ by z, the equation becomes

$$z^2 + 6z = 7.$$

From the latter they found that $z = 1$. So a root of the original equation is $x = \frac{1}{7}$.

The following example shows how the Babylonians formulated an algebraic problem and solved it.

EXAMPLE 5   If the (length of the) side of a square is subtracted from the (area of the) square, we get 14,30.

Solution. The statement of this example is equivalent to the following equation:

$$x^2 - x = 14,30.$$

We give in separate columns the Babylonian solution of this equation and its modern equivalent.

Babylonian	*Modern*
Take 1, the coefficient (of x).	Let $x - 1 = y$. Then $x - y = 1$ and $xy = 14,30$ (since $x^2 - x = x(x - 1)$).
Break 1 into two halves. $0;30 \times 0;30 = 0;15$ to add to $14,30$.	Let $x = a + 0;30$. Then $y = a - 0;30$. $(a + 0;30)(a - 0;30) = 14,30$. $a^2 = 14,30 + 0;15 = 14,30;15$.
And $14,30;15$ gives $29;30$ as a root.	$a = \sqrt{14,30;15} = 29;30$.
The $0;30$, which has been multiplied by itself, is added to $29;30$.	$x = 29;30 + 0;30$.
And 30 is (the side of) the square.	$x = 30$.

EXERCISES 2-5

1 Find in the Babylonian way two numbers that have a sum of 30 and a product of 104.

2 Find in the Babylonian way two numbers that have a difference of 3 and a product of 40. (Hint: Write the numbers as $a + \frac{3}{2}$ and $a - \frac{3}{2}$.)

3 See Example 3.
   (a) Find the other root of the quadratic equation.
   (b) Why did the Babylonians find only one root?
   (c) What are the additional roots in Examples 4 and 5?

4 Solve the following equations by use of a Babylonian procedure. Check your answers. Use modern procedures to give a second solution.
   (a) $x^2 + 4x = 21$
   (b) $x^2 - 2x = 4$
   (c) $x^2 + x = 7$

5 Solve $3x^2 + 4x = 4$ by use of the procedure of Example 4.

2-6 A BABYLONIAN TEXT

To become more familiar with the characteristics of Babylonian algebra, let us examine a problem as it appears on the Babylonian tablet shown in the photograph of Figure 2-2 and in the more legible drawing of Figure 2-3. The numbers in the margin of our translation correspond with the numbers in the margin of Figure 2-3. Items enclosed in brackets have been filled in by the translator.

Figure 2-2 Photograph of the side of the clay tablet (AO 8862) which contains the text discussed on pages 56 through 57. (From *Revue d'Assyriologie* 29 (1932).)

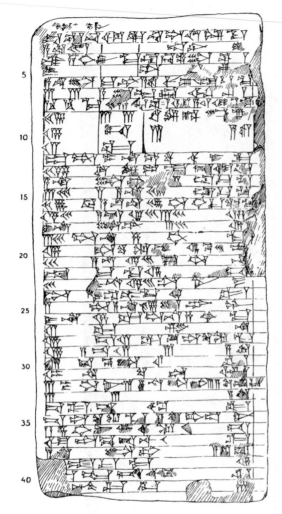

Figure 2-3 Copy of the contents of the clay tablet (AO 8862) pictured in Figure 2-2. (From O. Neugebauer, *Mathematische Keilschrifttexte*, Julius Springer, 1935-37.)

    Length, width. Length and width I have multiplied and thus formed the area. I have further added the excess of the length over the width to the area: 3,3. Further, I have added the length and the width: 27.

    Required: length, width, and area.

9   [Given:] 27 and 3,3, the sums.

10-11   [Result:] 15 length, 12 width, 3,0 area.

    You follow this method:

13-15 $27 + 3,3 = 3,30$

15-16 $2 + 27 = 29.$

16   Take the half of 29 [that is, 14;30].

17    $14;30 \times 14;30 = 3,30;15$

18-20 $3,30;15 - 3,30 = 0;15$

20-21   0;15 has 0;30 as square root.

21-22	14;30 + 0;30 = 15 length.
23-24	14;30 − 0;30 = 14 width.
25-26	Subtract the 2 you added to 27 from 14, the width:
27	12, the real width.
28	15, length, 12, width; I have multiplied.
29	15 × 12 = 3,0 area.
30-32	15 − 12 = 3
32-33	3,0 + 3 = 3,3.

We shall now try to understand the text by expressing it in modern notation, sentence by sentence.

<table>
<tr><td colspan="2" align="center">*Babylonian*</td><td align="center">*Modern*</td></tr>
<tr><td></td><td>Length, width.</td><td>Let the length be x, the width y.</td></tr>
<tr><td></td><td>Length and width I have multiplied and thus formed the area.</td><td>The area is then equal to xy.</td></tr>
<tr><td></td><td>I have further added the excess of the length over the width to the area; 3,3.</td><td>$x - y + xy = 3,3$ (1)</td></tr>
<tr><td></td><td>Further, I have added the length and the width: 27.</td><td>$x + y = 27$ (2)</td></tr>
<tr><td></td><td>You follow this method:</td><td>Add (1) and (2), thus:</td></tr>
<tr><td>13-15</td><td>27 + 3,3 = 3,30</td><td>$x + y + x - y + xy = 3,30$
$x(2 + y) = 3,30$ (3)</td></tr>
<tr><td>15-16</td><td>2 + 27 = 29</td><td>$2 + x + y = 29.$ (4)
Here a new variable is introduced,
$y' = y + 2.$
(This follows from lines 25-27 in the text.)
We now write (3) and (4) as
$xy' = 3,30$
$x + y' = 29.$</td></tr>
<tr><td>16</td><td>Take the half of 29, that is, 14;30.</td><td>Let $x = 14;30 + a$ and $y' = 14;30 - a$.
$(14;30)^2 - a^2 = 3,30$</td></tr>
<tr><td>17</td><td>14;30 × 14;30 = 3,30;15</td><td>$3,30;15 - a^2 = 3,30$</td></tr>
<tr><td>18-20</td><td>3,30;15 − 3,30 = 0;15</td><td>$a^2 = 3,30;15 - 3,30 = 0;15$</td></tr>
<tr><td>20-21</td><td>The square root of 0;15 is 0;30.</td><td>$a = \sqrt{0;15} = 0;30.$</td></tr>
<tr><td>21-22</td><td>14;30 + 0;30 = 15 length.</td><td>Hence, $x = 14;30 + 0;30 = 15$</td></tr>
<tr><td>23-24</td><td>14;30 − 0;30 = 14 width.</td><td>$y' = 14;30 - 0;30 = 14.$</td></tr>
<tr><td>25-26</td><td>Subtract the 2 you added to 27 from 14, the width:</td><td>$y' = y + 2$
$y = y' - 2 = 14 - 2.$</td></tr>
<tr><td>27</td><td>12, the real width.</td><td>So $y = 12.$</td></tr>
<tr><td>28-29</td><td>15, length, 12, width; I have multiplied.
15 × 12 = 3,0 area.</td><td>$x = 15, y = 12$, so $xy = 3,0.$</td></tr>
<tr><td></td><td></td><td>Now, to check:</td></tr>
<tr><td>30-32</td><td>15 − 12 = 3</td><td>$x - y = 3$</td></tr>
<tr><td>32-33</td><td>3,0 + 3 = 3,3.</td><td>$xy + x - y = 3,3.$</td></tr>
</table>

The foregoing text is characteristic of Babylonian algebra. The Babylonians followed the same line of thought as we do, but did not have anything approximating modern notation. They could therefore only express their procedures by means of numerical examples. That is why we find in their texts a great many problems that are all solved in the same way. The solutions nearly always terminate with the statement: "such is the procedure." Thus, it appears that these problems were intended to demonstrate a general method of solution.

The Babylonians were not used to working with abstract numbers as we are. They therefore talked of quantities, of what we sometimes term *denominate numbers*, numbers associated with units of measure (in the example: length, width, and area). Yet it does not appear that they had formulated a real geometrical problem, for in that case they probably would not have added length to area. Terms such as "length" only served to give a name to the unknown numbers, a process similar to what we often do in our arithmetic problems.

The preceding examples illustrate the fact that the Babylonians were aware of connections between geometry and algebra. Geometric terminology was added to give concreteness to the solution of an algebraic problem. However, their procedure of subtracting a length from an area shows that they had no objections to mixing dimensions.

In later centuries there was reluctance to mix "dimensions," that is, to combine numbers representing areas with those representing lengths or volumes. This reluctance is found from the Greek period through the work of the Italian algebraists of the sixteenth century and even up to the time of *Descartes* (1596-1650). Descartes, in his development of analytic geometry and the related theory of equations, did write and deal with expressions such as $3x^4 - 4x^3 + 5x^2 - 6x = 7$, where terms were combined even though their geometric counterparts would have been "squares" and "cubes," which are measured in different units. Although the development of the theory of equations was often helped by the use of geometric diagrams and concepts, it also was hindered by too-close ties to geometry. For example, there are no simple geometric counterparts for fourth- and higher-degree terms; as a result, when solving higher-degree equations, early algebraists either did not deal with such counterparts or resorted to unnecessarily awkward terms and processes.

2-7 BABYLONIAN GEOMETRY

The Babylonians had no formulated geometric theorems and proofs as we know them. Their problems were restricted to computation of the lengths of line segments and to computation of areas. From these computations it does appear, however, that the Babylonians were acquainted with several theorems: for example, the *Pythagorean theorem*.

EXAMPLE 1   (See Figure 2-4.) "A patu (beam?) 30 long (stands vertically against a wall). The upper end has shifted 6 lower. How far has the lower end moved?" The whole problem amounts to the calculation of one leg of a right triangle when the hypotenuse is 30 and the difference between the hypotenuse and the other leg is 6. The arithmetic computation shows that the Pythagorean theorem was applied. This computation was as follows.

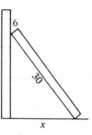
Figure 2–4

$$30 - 6 = 24, \qquad \text{so the height is 24.}$$
$$30^2 = 900$$
$$24^2 = 576$$
$$30^2 - 24^2 = 324.$$

Hence, $x = \sqrt{324} = 18$.

EXAMPLE 2   On a badly damaged clay tablet the diagram of Figure 2-5 is found. It can be completed easily into Figure 2-6.

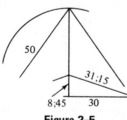
Figure 2–5

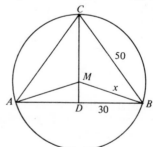
Figure 2–6

Apparently, the calculation of the radius of the circle circumscribed about the isosceles triangle ABC is required, given that $AB = 60$ and $AC = BC = 50$.

First, CD in $\triangle CBD$ can be calculated with the use of the Pythagorean theorem. It is found that $CD = 40$.

Suppose that $MB = x$; then $MD = 40 - x$. By applying the Pythagorean theorem to $\triangle MBD$, we see that

$$x^2 = (40 - x)^2 + 30^2$$
$$80x = 2500$$
$$x = 31\tfrac{1}{4}$$
$$MD = 40 - 31\tfrac{1}{4} = 8\tfrac{3}{4}.$$

Sexagesimally, the result reads

$$MB = 31;15, \qquad MD = 8;45.$$

The following example shows that the Babylonains had gone far in solving equations. It leads to the solution of three equations with three unknowns where only one of the equations is linear.

EXAMPLE 3   (See Figure 2-7.) We are given that the right triangle ABC is divided by $\overline{ED}$ into trapezoid $ECAD$ and triangle BED, and that $AC = 30$, area $ECAD$ — area $BED = 420$, and $BE - EC = 20$. We are required to calculate BE, EC, and ED.

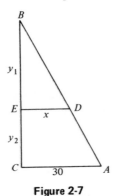

Figure 2-7

From the data we deduce the following two equations:

$$\tfrac{1}{2}y_2(x + 30) - \tfrac{1}{2}y_1 x = 420$$
$$y_1 - y_2 = 20.$$

To be able to calculate the three unknowns x, y_1, and y_2, a third equation is needed. This equation, which follows immediately from the similarity of $\triangle BED$ and $\triangle BCA$, is

$$\frac{y_1}{y_1 + y_2} = \frac{x}{30}.$$

The Babylonians solved this system of three equations. Try it yourself! You will then appreciate the high level of the Babylonian algebraic achievement.

These examples show that the Babylonians were acquainted with the following ideas: (1) the Pythagorean theorem; (2) that in an isosceles triangle, the line joining the vertex and the midpoint of the base is perpendicular to the base; (3) how to calculate the area of a triangle, a square, and a trapezoid; and (4) the proportionality of the corresponding sides of similar triangles. To what extent the Babylonians recognized these ideas explicitly, we can only conjecture.

2-8   APPROXIMATIONS TO π

The following problem concerns the calculation of the area of a circle. The translation is somewhat simplified.

EXAMPLE 1   "I have drawn the boundary of a city [the inner circle in Figure 2-8]. I do not know its length. I have walked 5 from the first circle away from the center in all directions and I have drawn a second boundary. The area between is 6,15. Find the diameter of the new and the old city."

The solution is as follows.

"Multiply the 5 of the increase with 3; you get 15.

"Take the inverse of 15 and multiply this by 6,15, the enclosed area; you get 25.

"Write this 25 twice.

"Add the 5, which you have walked, to the obtained result, and also subtract this 5 from it; you find 30 for the new city and 20 for the old city."

An explanation of the Babylonian solution follows. Let the radii of the new and the old city be represented by R and r, respectively. Then the area of the enclosed region is

$$A = \pi R^2 - \pi r^2 = \pi (R - r)(R + r).$$

Here, $A = 6,15$ and $R - r = 5$. In this problem the Babylonian uses 3 as an approximation to π. Hence,

$$6,15 = 3 \cdot 5(R + r).$$

We find $R + r$ by multiplying 6,15 by the inverse of 15, that is, by 0;4. Hence, we get

$$R + r = 0;4 \times 6,15 = 25.$$

By adding $R + r$ and $R - r$, we obtain

$$2R = 25 + 5 = 30,$$

the diameter of the new city. And by subtracting $R - r$ from $R + r$ we find

$$2r = 25 - 5 = 20,$$

the diameter of the old city.

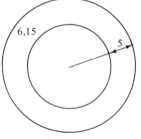

Figure 2-8

It is interesting to note that the value 3 for π, used in this example, also underlies the description of the dimensions of the bath for the use of priests in the temple built by Solomon (I Kings 7). "He then made the sea of cast metal; it was round in shape, the diameter from rim to rim being ten cubits; it stood five cubits high, and it took a line thirty cubits long to go round it."

In other problems, the Babylonian process for finding the area of a circle amounts to using the formula

$$A = \frac{1}{12}C^2,$$

where A is the area and C the circumference. It is easy to see that this relationship is exactly correct if 3 is used as a value for π. We know that

$$C = 2\pi r \quad \text{and} \quad A = \pi r^2,$$

from which it follows that

$$A = \pi\left(\frac{C}{2\pi}\right)^2 = \frac{C^2}{4\pi}.$$

If π were 3, this formula would become

$$A = \frac{C^2}{12}.$$

Besides the value 3 for π, the Babylonians occasionally used a better approximation, $3\frac{1}{8}$.

2-9   ANOTHER PROBLEM AND
A FAREWELL TO THE BABYLONIANS

The given examples clearly demonstrate that Babylonian mathematics was *not exclusively focused on applications*. A problem such as the calculation of the radius of the circumscribed circle of an isosceles triangle (Example 2 of Section 2-7) does not occur in daily life. Neither is Example 1 of Section 2-8 a problem from daily life, especially if we realize that Babylonian cities were square, not round.

If we compare Babylonian mathematics with Egyptian mathematics, we observe that Egyptian mathematics is more exclusively aimed at practical applications, whereas the Babylonians already show a beginning of a theoretical interest in mathematical problems.

Babylonian algebra has made good progress; rather complicated systems of equations can be solved. A handicap is the lack of a convenient literal notation. For this reason, general methods can only be explained by numerical examples.

In Babylonian geometry, no theorems are stated explicitly as such and none are proved. Babylonian geometrical problems only involve numerical calculations. However, we see evidence for the development of theoretical interests and for the use of interrelationships between algebra and geometry. On the other hand, such relationships often seem to be geometrically contrived to give a concrete basis for an algebraic type of problem. Further evidence for these theoretical interests is to be found in the tablet shown as Figure 2-1, with a discussion of which we shall conclude this chapter.

Pythagorean triples are sets of three whole numbers, such as $(3, 4, 5)$ and $(5, 12, 13)$ (in general: (a, b, c)), which satisfy the relation $a^2 + b^2 = c^2$. The followers of Pythagoras eventually solved the problem of finding procedures that give *all* such triples (there is an infinite number of them). Although the formula $a^2 + b^2 = c^2$ is reminiscent of the Pythagorean theorem for right triangles, the Pythagorean triple problem is essentially algebraic. However, the Babylonian tablet called Plimpton 322 (dating from between 1900 and 1600 B.C.) shows that the Babylonians had studied this problem much earlier. The tablet merely lists a series of Pythagorean triples, but the order in which they are listed makes us believe that the Babylonians had a general and systematic solution for the problem of finding Pythagorean triples.

EXERCISES 2-9

1 Show that if $u = 5$, then the formulas $a = u^2 - 1$, $b = 2u$, $c = u^2 + 1$ will give three numbers, a, b, and c, which satisfy the Pythagorean formula $a^2 + b^2 = c^2$.

2 The two middle columns of the tablet Plimpton 322 give numbers corresponding to b and c in the formula $a^2 + b^2 = c^2$. The first row gives the sexagesimal numbers 1,59 and 2,49. Find the replacement for a such that $a^2 + b^2 = c^2$ if $b = 1,59$ and $c = 2,49$.

3 Repeat Exercise 2; use the following values from the tablet:
   (a) fifth line: $b = 1,5$ and $c = 1,37$,
   (b) eleventh line: $b = 45$ and $c = 1,15$.

4 Calculate the circumference of a circle with diameter 10 if
   (a) $\pi = 3$
   (b) $\pi = 3\frac{1}{8}$
   (c) $\pi = 3.1416$
   Assume that the answer in (c) is correct. Calculate the error in the answer in (a). What percent of the correct answer is the error?

5 Repeat Exercise 4; this time use a diameter of 20.

6 Repeat Exercise 4; replace "circumference" by "area."

7 Repeat Exercise 4; use a diameter of 20 and replace "circumference" by "area."

8 On a Babylonian tablet, the following problem is solved: Given that the circumference of a circle is 60 and that the length of the sagitta $\overline{AB}$ of a segment is 2, calculate the length of the chord $\overline{CD}$ of the segment (see Figure 2-9). In solving this problem, use $\pi = 3$.

9 Solve the equations in Example 3, Section 2-7.

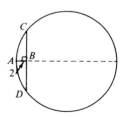

Figure 2-9

REFERENCES

For further data on Babylonian mathematics, see references 3, 4, 5, 6, 7 (Chapter 1), and

[16] Aaboe, Asger, *Episodes from the Early History of Mathematics*. New York: Random House, Inc., 1964 (Volume 13 of The New Mathematical Library).

[17] Vogel, Kurt, *Vorgriechische Mathematik*, Teil II. Hannover, West Germany: Hermann Schroedel Verlag KG, 1959.

3

THE BEGINNING
OF GREEK MATHEMATICS

3-1 THE EARLIEST RECORDS

The first mathematician who can be named is the Greek *Thales*, who lived about 600 B.C. From Thales to *Euclid* (300 B.C.) the chief workers in the field of mathematics were *Pythagoras* (540 B.C.) and *Hippocrates* (460 B.C.). *Plato* (360 B.C.) and *Aristotle* (340 B.C.), although primarily philosophers, had strong interests in and effects upon the development of mathematics. Their more mathematical contemporaries included *Archytas, Theodorus, Theaetetus, Eudemus,* and especially *Eudoxus* (360 B.C.) and *Menaechmus* (350 B.C.).

There is some uncertainty with respect to the exact dates of these mathematicians and philosophers. It is only as a result of scholarly detective work that we know much of their mathematics. Their original manuscripts have all been lost. Nevertheless, Plato's and Aristotle's writings and Euclid's *Elements* have been preserved through repeated copying. The oldest copies known date from the first centuries after Christ, that is, about 500 years after their date of origin. Inaccuracies and later additions resulted from the repeated copying, and it is only in the last century that these have been removed.

Probably very few copies were made of the earliest writings. How they were spread we do not know, but that they were appears from the fact that the writers influenced and opposed each other. A book trade seems actually to have arisen in the second half of the fifth century B.C.

In Plato's *Apology* it is said that Socrates speaks of the writings of the philosopher Anaxagoras (about 460 B.C.), and Socrates then remarks that occasionally the work can be bought on the market for one drachma. This story is set in the year 399 B.C.

We learn from the writings of Plato and Aristotle that they had studied the works of their predecessors, for in several places they refer to the older writers. Aristotle was especially interested in the progress of science in earlier years. He persuaded several of his pupils to write the history of a science. Eudemus (350 B.C.), for example, wrote a history of mathematics. This work has disappeared, but it is mentioned and cited by the later writers *Proclus* (about A.D. 450) and *Simplicius* (about A.D. 525). The oldest mathematics book that we know in its entirety is the *Elements*, by Euclid. Proclus wrote a commentary on Book I of the *Elements* (today Euclid's "Books" would be called "Chapters" or "Parts"). This commentary also gives some information about the geometry that was known before Euclid's time.

3-2   GREEK NUMERATION SYSTEMS

The last periods of the Egyptian and Babylonian empires are known as the Ptolemaic and Seleucid periods, respectively. They began about 323 B.C. with the death of Alexander the Great, king of Macedonia. Alexander's conquests extended from Greece to India and included Egypt and Babylonia, where he left the Greek generals Ptolemy and Seleucus in charge. Upon Alexander's death, they took over as kings. The period of Greek dominance and civilization, which extended from 600 B.C. to A.D. 500, overlapped the Egyptian-Babylonian period (3000-300 B.C.). Therefore, there was much opportunity for close contact between Greek culture and the earlier civilizations.

Within any one nation, languages and written symbols change with time. The old English of Chaucer and *Beowulf* seems almost a foreign language to us. Similarly, Egyptian writing progressed from hieroglyphic to hieratic. Numeration systems change and develop like languages and show the effects of the interaction of different cultures.

The Greeks used two systems of numeration. In earlier times they developed the *Herodianic system*. At a later time, they used the *Ionic system* to write their numerals.

The Herodianic system used the following symbols for numerals:

$$| = 1, \quad \Gamma = 5, \quad \Delta = 10, \quad H = 100, \quad X = 1000, \text{ and } M = 10,000.$$

The last five of these symbols are the initial letters of Greek number words. Several number words in modern languages reflect this system. The symbol Γ is an old form of the capital letter pi. It corresponds to our letter "p," and "pentagon," "pentameter," and "pentathlon" stem from the Greek word for five, *pente*. Similarly, delta, $\triangle$, stood for *deka* (10), which gives us "decagon" and "decameter" (10 meters). Words such as "decile," "decimal," "decimate," and "decimeter" (one tenth of a meter) came into English from the Latin *decem*, but also show their Greek origin. Eta, H, stood for *hekaton* (100) and leads to "hectare"; chi, X, stood for *kilioi* (1000) and leads to "kilometer" and "kilogram"; and mu, M, stood for *myrioi*, from which we get "myriad."

Numerals for other numbers were constructed from the above symbols additively by repeating them, such as

$$\triangle\triangle\triangle\triangle \ = 40,$$

multiplicatively by such devices as

$$\ulcorner^{\triangle} = 5 \cdot 10 = 50,$$

or

$$\ulcorner^{\mathsf{X}} = 5 \cdot 1000 = 5000,$$

and by more elaborate combinations, such as

$$\mathsf{X X \ulcorner^{\mathsf{H}} H \triangle \triangle \Gamma I} = 2626.$$

In the Ionic system, *the letters of the alphabet were used as digits.* The association of letters with digits was as follows:

1	2	3	4	5	6	7	8	9
α	β	γ	δ	ϵ	ς	ζ	η	θ

10	20	30	40	50	60	70	80	90
ι	κ	λ	μ	ν	ξ	o	π	ς

100	200	300	400	500	600	700	800	900
ρ	σ	τ	υ	φ	χ	ψ	ω	$\pi\!\wedge$

1000	2000	3000	4000	5000	6000	7000	8000	9000
$,\alpha$	$,\beta$	$,\gamma$	$,\delta$	$,\epsilon$	$,\varsigma$	$,\zeta$	$,\eta$	$,\theta$

A list of names for the Greek letters appears following the Preface. Since the classical Greek alphabet contained only 24 letters, three older letters were added to get the required 27 symbols. They were

vau, $\varsigma = 6$;   koppa, $\varsigma = 90$;   sampi, $\pi\!\wedge = 900$.

Two problems arose in this system: how to distinguish numerals from words, and how to write symbols for numbers larger than 999. The first problem was solved by drawing lines over numerals or by adding an accent at the end. The second problem was solved by several devices involving multiplicative ideas. A stroke before a numeral multiplied the number represented by that numeral by 1000; thus, $,\alpha = 1000$, $,\beta = 2000$, and so on. For still larger numbers, the M of the earlier, Herodianic, system was used with an Ionic symbol above it for a multiplier. Thus,

$$\overset{\beta}{M} = 2 \cdot 10{,}000 = 20{,}000.$$

EXAMPLES

$$\iota\gamma \ (= \overline{\iota\gamma} = \iota\gamma') = 13, \qquad \xi\epsilon = 65, \qquad \sigma\lambda\zeta = 237, \qquad \tau\alpha = 301,$$

$$,\gamma\psi\lambda\eta = 3738, \qquad ,\epsilon\eta = 5008, \qquad \overset{\lambda\eta}{M},\alpha\varphi o\delta = 381{,}574.$$

Under the influence of the Egyptians, the Greeks originally wrote fractions as a sum of unit fractions. At a later time they began writing a fraction as a pair of numerals. This was accomplished in several ways. Sometimes they wrote the numerator marked with a single accent, followed by the denominator written twice and both times marked with a double accent. For example,

$$\frac{2}{3} = \beta'\,\gamma''\,\gamma''.$$

In another representation, the Greeks wrote the denominator above the numerator but with no bar between them. Thus,

$$\frac{2}{3} = \frac{\gamma}{\beta}.$$

EXERCISES 3-2

1 Write the following numbers in Ionic notation:

   (a) 23   (b) 107   (c) 227

   (d) 8256   (e) 769,305   (f) $\frac{3}{5}$

   (g) $\frac{19}{21}$ (h) $\frac{328}{507}$

2 Write the following numbers in modern notation:

   (a) $\lambda\epsilon$ (b) $\kappa\alpha$ (c) $\phi\zeta\varsigma$

                 $\pi\epsilon$ $\tau\kappa\theta$

   (d) $,\epsilon\chi o\eta$ (e) $\overset{\pi\epsilon}{M},\varsigma\pi\gamma$ (f) $\overset{\tau\kappa\theta}{M},\delta\eta$

   (g) $\lambda'\mu\epsilon''\mu\epsilon''$ (h) $\delta'\theta''\theta''$ (i) $\frac{\mu\alpha}{\lambda\epsilon}$

   (j) $\frac{\omega\pi\gamma}{\lambda\zeta}$

3 Add:

   (a) $\kappa\alpha$ (b) $\quad ,\alpha\chi\xi\theta$

       $\upsilon\xi\alpha$

       $\underline{o\epsilon}$ $\overset{\epsilon}{\underline{M},\epsilon\phi\lambda\beta}$

4 Multiply $\phi\mu\eta$ by $\iota\eta$.

5 Write the numbers in Exercise 1(a) through (d) in Herodianic notation.

3-3 THALES AND HIS IMPORTANCE TO MATHEMATICS

The earliest center of Greek civilization was actually in the colonies (Miletus, Ephesus) on the western coast of Asia Minor, which had developed more quickly than the home country. The *Iliad* and the *Odyssey* had their origin there. About 600 years before Christ, these Greek colonies were very prosperous. Trading and seafaring were the most important means of subsistence. The colonies had relations with Babylon (by land) and with Egypt (by sea). Nautical science and astronomy were of much importance. Ships had to be built, and therefore industry was developed.

Through their voyages, the Greek colonists became acquainted with disciplines they had not known before. Remarkably, the Greeks showed a new kind of interest in them right from the beginning. They were concerned with the questions about the how and the why of things, and they were not primarily interested in practical applications. The man who took the first steps on the road to our modern science is Thales of Miletus (624-547 B.C.), known as one of the "seven wise men."

Thales was a merchant, and it is said that he traveled much and spent a long time in Egypt and Babylon. In Egypt he astonished King Amasis by determining the height of a pyramid from the length of its shadow. According to one story, he put a stick vertically into the ground and waited until the lengths of the stick and its shadow were equal; then the height of the pyramid was equal to the length of its shadow, which could be measured. It is also said that he predicted a solar eclipse and that he did this on the basis of knowledge he had acquired from the Babylonians. The latter had occupied themselves with the prediction of eclipses of the sun and the moon as early as 700 B.C., according to letters from the Assyrian court astrologers. Although probably these stories about Thales have to be considered as legends, they nevertheless throw light on his fame and on the mathematical-scientific interests of his time.

Thales' importance for mathematics lies chiefly in that *he sought a logical foundation of geometrical theorems*. He did not produce a complete system of theorems and demonstrations as we know them today, but he was the first to venture in this direction. The theorems he pondered are quite simple. According to later writers, they are supposed to have been the following five:

1. The angles at the base of an isosceles triangle are congruent.
2. If two straight lines cut one another, the vertical angles are congruent.

3. Two triangles are congruent if they have two angles and the included side of one congruent to two angles and the included side of the other.
4. A circle is bisected by a diameter.
5. An angle inscribed in a semicircle is a right angle.

We are told that Thales used Theorem 3 in finding the distance of a ship from the shore. We do not know with certainty what this method was, but Figure 3-1 suggests one possibility. Suppose that the ship is at point S and the observer on the shore is at point W. The required distance is SW. The observer walks an appropriate distance, WM, along the shore at right angles to $\overline{SW}$ and puts a stick in the sand at M. He continues to walk in the same direction for a distance MP equal to $\underline{WM}$. He is then at point P. Finally, he walks inland, at right angles to $\overline{WP}$, until he sees the ship at S and the stick at M in a straight line. He is then at Q, and PQ is the required distance, which can be measured.

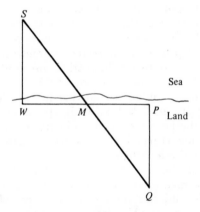

Figure 3-1

According to Proclus, Thales brought geometry to Greece from Egypt and proved Theorem 4 himself. On the other hand, it is said that Thales discovered Theorem 2 but did not prove it.

How might he have given such proofs? It is difficult to believe that Thales, the first to think along these lines, started immediately with basic ideas and axioms, as we do. If he did not have a set of axioms to begin with, his demonstrations must have been of an intuitive character, not proofs in a modern sense. He might have proved Theorem 4, for instance, by folding a circle along a diameter.

There are important points of difference between the geometry of Thales and the geometry of the Egyptians and the Babylonians. *Thales was the first to formulate properties of figures as general statements.* It had not occurred to the Egyptians and Babylonians to mention them explicitly. It appears that Thales was interested in geometric figures as

such. Perhaps he felt the need to base even apparently obvious properties upon logical arguments. Furthermore, the properties that he found
interesting are of a different character than those with which the Babylonians and Egyptians were occupied; their problems always involved
calculating something (even though, with the Babylonians, it was not
always something practical they calculated). Thales' interest, on the
other hand, was directed to the geometric properties of the figures.

EXERCISES 3-3

1 See Figure 3-1. How could Thales have proved that $PQ = SW$?

2 Which of the Theorems 1 through 5 are used in the proof in Exercise 1?

3 Give a modern proof of Theorem 1.

4 Give a modern proof of Theorem 2.

5 Consider the determination of the height of a pyramid by Thales. Assume that
   light rays from the sun are parallel.
   (a) Draw a diagram for the problem.
   (b) What kind of triangles are formed?
   (c) What theorems are needed to prove Thales' method valid?
   (d) How could he have used a shadow method at any time of the day?

6 There are many legends about Thales as a merchant, philosopher, and politician.
   Look these up in an encyclopedia or in the references at the end of the chapter.

3-4 PYTHAGORAS AND THE PYTHAGOREANS

Pythagoras lived about 570-500 B.C. He probably traveled in Egypt as
well as in Asia Minor and thus took note of both the Egyptian and the
Babylonian civilizations. Later, he settled in the southern part of the
Italian peninsula. He opened a school there, where he propounded his
ideas mainly to the upper classes of the population. He selected from
those who attended his lectures the people who showed particular interest and talent. They were initiated into the deeper secrets of the
Pythagorean Order, which used a five-pointed star (pentagram) as its
mystical symbol. In fact, much of the Pythagorean doctrine was mystical; it also contained many ethical beliefs, but all had a mathematical
foundation. Many of the teachings of the Pythagoreans are lost to us,
because the members of the order were bound by an oath not to reveal
the teachings of the master. We do know that the Pythagorean Archytas
divided mathematics into four parts: *music, arithmetic, astronomy*, and
geometry. These subjects, called the *quadrivium*, were later adopted by
Plato and Aristotle and became the school curriculum for centuries—
in fact, up until the Renaissance. We shall present enough of the Pytha-

gorean concepts relative to each of the elements of the quadrivium to show how number was fundamental to all of them.

The legend of how Pythagoras discovered numbers in music will help us to understand how mathematics became mixed with mysticism.

3-5 THE PYTHAGOREANS AND MUSIC

The Roman *Boethius* (sixth century A.D.) tells the following. On passing a blacksmith shop, Pythagoras was struck by the fact that the sounds caused by the beating of different hammers on the anvil formed a fairly musical whole. He entered the smithy. Could the phenomenon be caused by the strength of the men? He asked them to exchange hammers. The musical effect was the same. So the hammers, not the men, were the cause. He asked the smith to let him weigh the hammers. The weights were in the ratio of the numbers 12, 9, 8, and 6. The weight of the fifth hammer did not have a simple ratio to any other. The fifth hammer was now left out and the musical effect appeared to be increased. The musical intervals that Pythagoras heard were those which are called the fourth, the fifth, and the octave, counted from the note caused by the heaviest hammer.

We cannot tell what is historic in this legend. At worst, there may be little truth in it, for the supposed result does not agree with the known relation between the weight of the hammers and the pitch of the sound. In any case, this legend is typical of the Pythagorean interest in the concept of number.

Boethius further tells us that Pythagoras continued his experiments and investigated the relation between the length of a vibrating string and the musical tone it produced. If a string was shortened to $\frac{3}{4}$ of its original length, then what is called the fourth of the original tone was heard; if shortened to $\frac{2}{3}$, the fifth was heard; and if shortened to $\frac{1}{2}$, the octave. If the original length of the string is equal to 12, then the fourth is heard when the string is shortened to length 9, the fifth when it is shortened to length 8, and the octave when it is shortened to length 6. This part of the story is probably closer to reality.

The perception of the relationship between the length of a string and the pitch of a tone is *the oldest example in history of a natural law found empirically*. It must have been extremely surprising at that time to discover a relation between things as different as musical tones and ratios of numbers. It is understandable that the Pythagoreans asked: What is the cause, and what is the effect? They ended by assuming that the cause was to be found in the ratios of numbers. Thus, to them, harmony became a property of numbers; musical harmony occurred because the numbers 6, 8, 9, and 12 were harmonic numbers.

To Pythagoras, the relation thus found was a proof that *number plays a part in the things with which we are in daily contact.* In the course of history it has happened several times that people were so much impressed and surprised by an original point of view that they lost sight of its special nature and believed that they could apply the new concept everywhere. This may also be said of the Pythagoreans. *Number was taken for that which is fundamental in all natural science.* They thought that all knowledge could be reduced to relations between numbers. This led to all kinds of speculations about numbers on the one hand and to great interest in mathematics on the other.

The numbers 6, 8, 9, and 12 appeared to have not only an acoustical but also an arithmetical peculiarity. For the number 9 is exactly the arithmetic mean of 6 and 12, and 8 is the harmonic mean of 6 and 12. That is

$$9 = \frac{1}{2}(6 + 12) \quad \text{and} \quad \frac{1}{8} = \frac{1}{2}\left(\frac{1}{6} + \frac{1}{12}\right).$$

Further, 6 is to 8 as 9 is to 12, and 6 is to 9 as 8 is to 12, or

$$\frac{6}{8} = \frac{9}{12} \quad \text{and} \quad \frac{6}{9} = \frac{8}{12}.$$

In music, this means that just as strings of length 12 and 9, also strings of length 8 and 6 produce a tone and its fourth, whereas strings of length 9 and 6 produce a tone and its fifth.

The *arithmetic mean* of two numbers was defined by the Pythagoreans as a number such that the mean exceeds the first number by the same amount as the second number exceeds the mean. Thus, 9 is the arithmetic mean of 6 and 12 because $9 - 6 = 12 - 9$. In other words, the arithmetic mean of two numbers can be defined as one half of their sum. Thus, the arithmetic mean of 6 and 12 is

$$\frac{6 + 12}{2}.$$

The *geometric mean* of two numbers was defined as a number such that the difference of the geometric mean and the first number divided by the difference of the second number and the geometric mean is equal to the first number divided by the geometric mean. Thus, 6 is the geometric mean of 2 and 18 because

$$\frac{6 - 2}{18 - 6} = \frac{2}{6}.$$

An equivalent modern definition is: The geometric mean of two numbers is the square root of their product. Thus,

$$6 = \sqrt{2 \cdot 18}.$$

The *harmonic mean* of two numbers was defined as a number such that the difference of the harmonic mean and the first number divided by the difference of the second number and the harmonic mean is equal to the first number divided by the second number. Thus, 8 is the harmonic mean of 6 and 12 because

$$\frac{8 - 6}{12 - 8} = \frac{6}{12}.$$

An equivalent modern definition of the harmonic mean, H, of two numbers, a and b, is

$$\frac{1}{H} = \frac{\dfrac{1}{a} + \dfrac{1}{b}}{2}.$$

The numbers 6, 8, and 12 showed particular properties in geometry as well as in arithmetic and music. A cube has 6 faces, 8 vertices, and 12 edges. Therefore, according to the Pythagoreans, the cube was a *harmonic body*. The reasoning is curious: In nature a musical harmony is found; it appears to be related to the numbers 6, 8, and 12. On this account *the harmonic character is regarded as a property of these numbers. So when these numbers occur elsewhere* (in this case with the cube), *there, too, the harmony peculiar to these numbers must occur.* Hence, the cube is a harmonic body.

Thus, number provided not only a theory for music but also a connection between music and geometry. Many additional connections between numbers and geometry will be discussed in Section 3-6.

EXERCISES 3-5

1   Find the arithmetic mean and the harmonic mean of each of the following pairs of numbers:
   (a)   7, 11   (b)   23, 72
   (c) $-3, -3$ (d)   4.5, 6.8

2   Does the pair $-3, +3$ have
   (a)   an arithmetic mean?
   (b)   a harmonic mean?

3   In statistics, the arithmetic mean (or average) of several observations is defined to be the sum of the numbers divided by the number of observations. Find the arithmetic mean of
   (a)   7, 9, 16, 25, 32, 50
   (b) -2, 12, 72, 108

4   The harmonic mean of a set of numbers is defined to be the reciprocal of the average of their reciprocals. Thus, the harmonic mean, H, of 3, 6, 12, and 24 is defined to be

$$H = \cfrac{1}{\cfrac{\frac{1}{3} + \frac{1}{6} + \frac{1}{12} + \frac{1}{24}}{4}} = \frac{4}{\frac{8 + 4 + 2 + 1}{24}} = \frac{4 \cdot 24}{15} = \frac{32}{5} = \frac{4 \cdot 24}{15} = \frac{32}{5}.$$

Find the harmonic mean of
(a) 3, 6, 12
(b) 6, 12, 24
(c) 5, 6, 7, 9

5 Show that the two definitions given in the text of the arithmetic mean, M, of two numbers a and b are equivalent. That is, show that

$$a - M = M - b \quad \text{if and only if} \quad M = \frac{a + b}{2}.$$

6 Show that the two definitions given in the text of the harmonic mean, H, of two numbers a and b are equivalent. That is, show that

$$\frac{H - a}{b - H} = \frac{a}{b} \quad \text{if and only if} \quad \frac{1}{H} = \frac{\frac{1}{a} + \frac{1}{b}}{2}.$$

7 Show that the harmonic mean, H, of a and b is given by the formula

$$H = \frac{2ab}{a + b}.$$

8 Show that the two definitions given in the text of the geometric mean, G, of two numbers a and b are equivalent. That is, show that

$$\frac{G - a}{b - G} = \frac{a}{G} \quad \text{if and only if} \quad G = \sqrt{a \cdot b}.$$

(Note: The Greeks did not use negative numbers.)

3-6 PYTHAGOREAN *ARITHMETICA*

In the United States, the word "arithmetic" usually refers to the computational procedures, or algorithms, for real numbers. In fact, arithmetic is commonly viewed as also including the solution of problems involving ratios, proportions, decimal fractions, and percent. This computational aspect of mathematics was called *logistica* (a word related to our "logistics") by the ancient Greeks. To them, *arithmetica* was a study of the abstract mathematical properties of numbers, chiefly the natural numbers, that is, the positive integers, or counting numbers. *Arithmetica* was the concern of philosophers and gentlemen of leisure; *logistica* was the concern of merchants and slaves. In the United States, the *arithmetica* of the ancient Greeks is called *number theory*. The theory of *figurate numbers* was one of their earliest and most extended developments in this field. It also shows interrelationships between *arithmetica* and geometry.

It seemed natural to the Pythagoreans to associate a point with the number 1; three points, or a triangle, with the number 3; and so on. This led to the following series of connected figures and numbers:

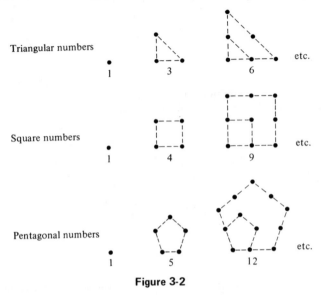

Figure 3-2

Each successive diagram (after the second) for a figurate number is constructed as follows:

1. Given a figurate number.
2. Connect the successive dots on the boundary of the dot polygon.
3. Choose one vertex of the polygon and extend the two edges meeting in this vertex.
4. Add one dot on each of these extensions.
5. Draw a regular polygon on these two extended sides.
6. Place on each side of the new polygon a number of dots equal to the number of dots on the extended sides in step 4. Then the figurate number consists of the totality of the dots.

The procedure is illustrated in Figure 3-3 for the construction of the fourth triangular number, given the third.

There is still another way in which a relation between the numbers and the geometrical figures was constructed. We mentioned already that a *point* was associated with the number 1. A *straight line* is determined by two points. Therefore, a straight line was associated with the number 2. Continuing in this way, a *plane* was associated with the number 3 and, finally, *space* with the number 4. In this way, the numbers 1, 2, 3, and 4 became the four fundamental numbers for geometry.

Square numbers and Pythagorean triples (mentioned in Chapter 2

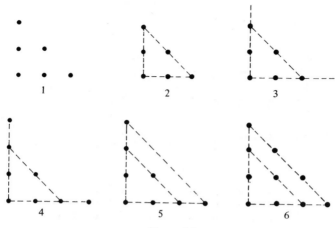

Figure 3-3

in connection with the Babylonian tablet Plimpton 322) have an interesting relationship. The Pythagoreans were familiar with the formula

$$\left(\frac{m^2 + 1}{2}\right)^2 = \left(\frac{m^2 - 1}{2}\right)^2 + m^2, \tag{1}$$

where m is an odd natural number. The equality may be checked readily. The formula was used as follows. Let m be an odd integer, for example, 11. Then

$$\frac{m^2 + 1}{2} = 61, \qquad \frac{m^2 - 1}{2} = 60,$$

and, since $$61^2 = 60^2 + 11^2,$$

$(60, 11, 61)$ is a Pythagorean triple.

The following derivation of formula (1) is plausible, considering the interest of the Pythagoreans in figurate numbers. The diagrams for square numbers show that to build from the first square (a single dot, representing 1) to the second square (representing 4) we added $1 + 1 + 1$ dots to the first. By adding $2 + 2 + 1$ dots to the figure representing 4, we arrived at the figure representing 9, the third square number. Figure 3-4 shows that to get from the nth square number to the $(n + 1)$th square number, we must add a row of n dots at the bottom, a column of n dots at the right, and a single dot "in the corner." Thus, the $(n + 1)$th square number is $2n + 1$ more than the nth square number; that is,

$$(n + 1)^2 = n^2 + (2n + 1). \tag{2}$$

(The reader recognizes the formula for the square of the binomial $n + 1$.)

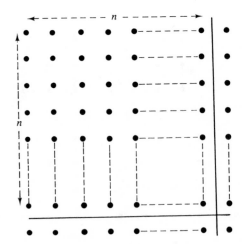

Figure 3-4

We observe that in formula (2), if $2n + 1$ is a square number, the conditions for a Pythagorean triple are present. Thus, if $2n + 1 = m^2$, then

$$n = \frac{m^2 - 1}{2}, \qquad n + 1 = \frac{m^2 + 1}{2},$$

and the formula

$$(n + 1)^2 = n^2 + (2n + 1)$$

becomes

$$\left(\frac{m^2 + 1}{2}\right)^2 = \left(\frac{m^2 - 1}{2}\right)^2 + m^2,$$

which is formula (1).

Formula (1) is, of course, true for every natural number m. But only odd values of m give Pythagorean triples (see the example for $m = 11$, giving the triple (60, 11, 61)). Even values of m lead to fractions, and *arithmetica* was restricted to integers. A later formula, which follows directly from formula (1), was

$$(m^2 + 1)^2 = (m^2 - 1)^2 + (2m)^2. \tag{3}$$

It works for any natural number m but gives the same Pythagorean triples as formula (1). Formula (3) is attributed to Plato. Check that for $m = 2, 3$, and 4, formula (3) gives the triples (3, 4, 5), (6, 8, 10), and (8, 15, 17). It does, in fact, produce an infinite number of triples, a different triple for each natural number substituted for m. The reader may wonder if there exist any triples that are not produced by this formula. We note that $7^2 + 24^2 = 25^2$, and we observe that 24 and 25 differ by 1, whereas the numbers $m^2 - 1$ and $m^2 + 1$ always differ by 2. Hence, the triple $(7, 24, 25)$ cannot be obtained from formula (3).

Euclid, in the *Elements*, gives a completely general and systematic solution to the problem of constructing Pythagorean triples. The procedure is as follows. If u and v are integers and if

$$x = u^2 - v^2, \qquad y = 2uv, \qquad z = u^2 + v^2,$$

then x, y, and z are integers such that

$$x^2 + y^2 = z^2.$$

For example, if $u = 3$ and $v = 2$, these formulas give the triple $(5, 12, 13)$, and if $u = 4$ and $v = 3$, they give the triple $(7, 24, 25)$. It can be shown that *all* possible Pythagorean triples can be produced by these formulas.

When the problem of generating Pythagorean triples was solved, many related problems were suggested to later mathematicians. One of these mathematicians was a Frenchman, *Pierre de Fermat* (1601-1665), who considered a generalization of Pythagorean triples. He stated that there are no natural numbers x, y, z, and n such that

$$x^n + y^n = z^n$$

if $n > 2$. For example, there are no natural numbers x, y, and z such that

$$x^3 + y^3 = z^3.$$

His statement appears as a remark written in the margin of a Latin translation of Diophantus' *Arithmetica*, along with a note that the margin was too small to contain the proof. Since that time, no one has been able to find either a proof of Fermat's theorem or a counterexample. In 1908 a prize of 100,000 marks was offered to anyone who would give a complete proof. No one has been awarded the prize, but it would now be worthless. However, much mathematics was developed in efforts to prove the theorem.

Let us now turn to another aspect of the Pythagorean *arithmetica*. The Pythagoreans devoted considerable attention to the investigation of the properties of individual numbers or classes of numbers. Odd, even, prime, and composite numbers were studied, and their properties were used in many ways. The Pythagoreans also searched for what they called *perfect numbers*, that is, numbers equal to the sum of their proper divisors. For example, 6 is a perfect number, because $6 = 3 + 2 + 1$ and 3, 2, and 1 are all the divisors of 6 that are different from 6. Other perfect numbers are 28 and 496. The Pythagoreans also looked for *friendly numbers*, that is, pairs of numbers which have the property that each number is equal to the sum of the proper divisors of the other (for example, 284 and 220).

The most developed part of the *arithmetica* was the *theory of even and odd*. Among others, the following theorems were known:

1. The sum of two even numbers is even.
2. The product of two odd numbers is odd.
3. When an odd number divides an even number, it also divides its half.

As we will soon see (Section 3-10), this theory, associated with the Pythagorean theorem, led the Pythagoreans to discover the existence of irrational numbers.

The search for perfect numbers has been carried on up to the present day and has led to the development of much mathematics, as well as to some still unsolved problems. The last theorem in Book IX of Euclid's *Elements* is equivalent to: *If*

$$2^n - 1 \tag{4}$$

is a prime, then

$$2^{n-1} (2^n - 1) \tag{5}$$

is a perfect number. For example, for $n = 2$,

$$2^2 - 1 = 3$$

is a prime, and hence

$$2^{2-1} (2^2 - 1)$$

is a perfect number. In fact, it is 6, the previously mentioned example. On the other hand, replacing n in formula (5) by 4 does not give a perfect number, since

$$2^4 - 1 = 15$$

(from formula (4)) is not a prime.

The converse of this theorem is a famous unproved conjecture: *Every perfect number is of the form*

$$2^{n-1} (2^n - 1),$$

with $2^n - 1$ a prime. Leonhard Euler (1707-1783), a Swiss mathematician, proved that all *even* perfect numbers are of this form. No one has ever found an *odd* perfect number, nor has it been proved that there is no odd perfect number.

EXERCISES 3-6

1 Draw figures for the fifth triangular number, the fifth square number, and the fourth pentagonal number, and show that they represent 15, 25, and 22, respectively.

2 The formula for the nth square number, S_n, is $S_n = n^2$. Find a formula for T_n, the nth triangular number.

3 Find a formula for P_n, the nth pentagonal number. (Hint: The nth pentagonal number can be shown to be equal to n plus three times the $(n-1)$th triangular number.)

4 Oblong numbers are represented by dot rectangles the length of which exceeds the width by one dot. The first two oblong numbers are $2 = 2 \times 1$ and $6 = 3 \times 2$.
   (a) Draw diagrams for these two and the next two oblong numbers.
   (b) Show by means of a figure that the nth oblong number is twice the nth triangular number.
   (c) Prove algebraically that the nth oblong number is twice the nth triangular number. (Hint: Use the formula for T_n, found in Exercise 2.)

5 Prove the formula $S_n = T_{n-1} + T_n$
   (a) by using a diagram,
   (b) algebraically.

6 Proclus ascribes to Pythagoras himself the following procedure for obtaining Pythagorean triples: Choose a natural number n and let $x = 2n + 1$, $y = 2n^2 + 2n$, and $z = 2n^2 + 2n + 1$.
   (a) Show that for each of $n = 1$, $n = 2$, $n = 3$, the procedure gives Pythagorean triples.
   (b) Prove that the procedure *always* produces a set of Pythagorean triples.
   (c) Show that the procedure will not produce all Pythagorean triples.

7 (a) Check that if $n \leqslant 10$, the sum of the first n odd numbers is a perfect square, n^2.
   (b) Prove that for every natural number n, the sum of the first n odd numbers is equal to n^2.

8 See Euclid's formula for perfect numbers (formula (5)). Show that $n = 3$ leads to a perfect number and that $n = 6$ does not.

9 See formula (5). Estimate the number of digits in the perfect numbers corresponding to
   (a) $n = 7$
   (b) $n = 17$
   (c) $n = 127$
   (Hint: The use of logarithms will help.)

10 (a) Prove the theorems about even and odd numbers mentioned on page 80.
   (b) Conjecture four additional theorems and prove or disprove them.

3-7 PYTHAGOREAN NUMEROLOGY

We recall the Pythagorean disposition to ascribe nonmathematical qualities to numbers, such as friendly and perfect. We must suppress the
tendency to label this disposition as naive. On the contrary, the speculations of the Pythagoreans about numbers not only had an intellectual
nature but were also permeated by a mystical significance. *Number is
the essence of things. As such, it has magical force.* Pythagoras was
probably influenced in this direction by the Babylonian culture with
which he came into contact when traveling in Asia Minor. A later
Pythagorean, *Philolaus* (about 450 B.C.), wrote: "And really everything
that is known has a number. For it is impossible that without it anything can be known or understood by reason. The One is the foundation
of everything."

Recalling the harmonic character of the cube, discussed previously
(page 74), Philolaus' statement shows again that there was a profound
meaning in assigning numbers to solids. These numbers were considered
the essential element. In a certain sense, the solid could be considered
as the manifestation of this element.

The mystical power granted to a number is even more clearly
shown by what Philolaus wrote about the number 10. The Greeks used
the decimal system. This in itself was an excellent reason to give the
number 10 a special importance. It is the smallest number that "contains" as many composite numbers as noncomposite numbers (the composite numbers are 4, 6, 8, 9, and 10; the noncomposite numbers are 1,
and the primes 2, 3, 5, and 7). Moreover, it is the sum of the numbers 1,
2, 3, and 4, the four numbers that were fundamental in geometry.
This remarkable number must therefore have very special properties,
also outside mathematics. That is why Philolaus wrote:

> The activity and the essence of the number must be measured by the
> power contained in the notion of 10. For this (power) is great, all-embracing,
> all-accomplishing, and is the fundament and guide of the divine and heavenly
> life as well as of human life . . . .
> Without this (power) everything is without limit, indistinct and vague.
> For the nature of number is to be informative, guiding and instructive
> for anybody in everything that is subject to doubt and that is unknown.
> For nothing about things would be comprehensible to anybody, neither
> of things in themselves, nor of one in relation to the other, if number and
> its essence were nonexistent . . . .
> One cannot only observe in the actions of demigods and gods the essence
> of the number and the power operative in it, but also everywhere, in all
> actions and words of men and in all branches of handicraft and in music.
> The essence of number, like harmony, does not allow misunderstanding,
> for this is strange to it. Deception and envy are inherent to the unbounded,

unknowable, and unreasonable. . . . Truth, however, is inherent in the nature
of number and inbred in it.

After this quotation it is not surprising to find that number was
considered by the Pythagoreans as the fundamental element of many
domains of nature and ethics. For example, the odd numbers were
called *masculine* and the even numbers *feminine*. The sum of the first
odd and the first even number, 5 (= 2 + 3; here, 1, being the primal
foundation of all numbers, was not regarded as a number), was the
symbol of marriage. A square number symbolized *justice* (render like
for like). There were also less understandable associations: 6 was the
number of the *soul*, 7 of *understanding* and *health*, 8 of *love* and
friendship.

3-8 PYTHAGOREAN ASTRONOMY

Speculations about numbers were even used as a basis for scientific
theories. This happened in the science of astronomy. According to the
Pythagoreans, the universe was built up of concentric spheres. Beginning
with the outer sphere, they were: the sphere of the fixed stars; the
spheres of each of the five planets; and the spheres of the sun, the
moon, and the earth. The sphere of the fixed stars was considered as
one heavenly body. There were then altogether nine heavenly bodies.
They all rotated about a common center, the central fire, where Zeus,
the governing power of the universe, resided. But it was inconceivable
that the number of heavenly bodies would be nine when the most
important number was 10. So there must be 10 heavenly bodies. The
Pythagoreans indeed assumed the existence of a tenth heavenly body,
the "counterearth." But there had to be a reason why this body was
not seen. Therefore, they supposed that the earth moved in such a way
that the inhabited side was always turned away from the central fire
and that the counterearth was diametrically placed relative to the earth
or between the earth and the central fire. For these reasons the counterearth could not be seen.

Also the distances of the heavenly bodies to the central fire were
subject to the laws of number. The numerical ratios that caused harmony in music must also bring about a harmonic structure in the universe. The ratios of the distances must therefore be harmonic. The
Pythagoreans went even further, for they assumed that the ratios of the
dimensions of the universe brought about corresponding musical tones.
The Pythagoreans made this assumption plausible by pointing out that
on earth, also, an object that rushes past causes a sound. Accordingly,
the motion of the heavenly bodies caused sounds. This heavenly music,
the harmony of the spheres, would not be noticed by men, however,

because it continued uninterruptedly. Only an interruption would be observed.

So we see that observational experience (for example, with vibrating strings) supported the mysticism of numbers. In the connections observed by Pythagoras between numbers and musical intervals, an argument was found supporting the correctness of the Pythagorean views with regard to numerology. On the other hand, starting from numerical mysticism, the Pythagoreans arrived at natural laws; that is, numerical mysticism led to a study of nature in which number speculations were accepted as a foundation for natural laws.

3-9   PYTHAGOREAN GEOMETRY

We noted earlier the connections between *arithmetica* and geometry embedded in the study of figurate numbers. The relationship between the numbers 1, 2, 3, and 4 and the geometrical notions point, line, plane, and space were also mentioned. However, we have not yet considered the development of geometry.

It is not known with certainty which of the early Greek geometrical discoveries should be ascribed to Pythagoras himself and which to his followers. We do know, however, that the Pythagoreans studied geometry apart from the just mentioned relations between geometry and numbers. In the fifth century B.C., plane geometry began to develop into the form in which we know it today. Geometry no longer consisted of scattered properties, the truth of which was clear on the ground of observation alone. A system of theorems was developed. Each theorem was proved on the strength of the preceding theorems. We know only a few details. It is certain that at this time the theorem of Pythagoras belonged to the set of theorems that had been proved.

The Pythagoreans apparently studied parallel lines, proved that the sum of the angles of a triangle is equal to the sum of two right angles, and developed the method of application of areas, which led to the geometric algebra to be found in the work of Euclid. They possibly were also acquainted with the five regular polyhedra, later called *Platonic solids*. Perhaps their most important accomplishment, however, was the discovery of irrational numbers, which occurred in a geometric setting.

3-10   INCOMMENSURABLE SEGMENTS AND
IRRATIONAL NUMBERS

Recall for a moment the difference between rational and irrational numbers. The *rational numbers* are the fractional numbers and the integers. (The Greeks did not have knowledge of negative numbers.)

A fractional number can be represented by the quotient a/b, where a and b are integers, $b \neq 0$. An integer a can also be represented as a quotient, $a/1$. Hence, *every rational number can be represented as the quotient of two integers*. Examples of rational numbers written in this way are

$$\frac{2}{3}, \ \frac{9}{5}, \ \frac{4}{14}, \ \frac{5}{1}, \ \frac{0}{2}.$$

The irrational numbers are real numbers that cannot be represented as a quotient of two integers. Examples of irrational numbers are

$$\sqrt{2}, \quad \sqrt[3]{5}, \quad \log_{10} 3, \quad \pi.$$

To show that $\sqrt{2}$ is irrational, we shall use the Pythagorean theorem: *In a right triangle the square described on the hypotenuse is equal in area to the sum of the squares described on the sides.* If we apply this theorem to a right isosceles triangle, as in Figure 3-5, we find that the square on the hypotenuse is equal in area to two times the square on a side. Since

$$AB^2 = AC^2 + BC^2$$

and $BC = AC,$

we have $AB^2 = 2AC^2.$

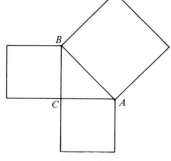

Figure 3-5

What is the ratio of AB to AC? The Pythagoreans tried to find two natural numbers proportional to AB and AC. Suppose that p and q are two such numbers and that they have no factor in common. Then we have

$$\frac{AB}{AC} = \frac{p}{q},$$

where p and q are not both even. Since

$$AB^2 = 2AC^2$$

we have $$\frac{AB^2}{AC^2} = 2,$$

and hence $$\frac{p^2}{q^2} = 2,$$

or $$p^2 = 2q^2.$$

So p^2 is an even number. Then, also, p is even (for "odd times odd is odd"). Hence, p can be replaced by $2r$ (in which r, again, is a natural number). Then

$$4r^2 = 2q^2$$

and hence

$$2r^2 = q^2.$$

From this it follows that q^2 is an even number and hence that q is even. Since p and q are not both even, and we already found that p is even, q must be odd. But a number cannot be both even and odd; therefore, our assumption (that AB and AC are proportional to two natural numbers) leads to a contradiction. Hence, *it is not possible that the hypotenuse and a side of an isosceles right triangle are in the ratio of two natural numbers.*

As a particular case, if AC is equal to 1, we have

$$AB = \frac{AB}{1} = \frac{AB}{AC},$$

which cannot be written as a quotient of two integers. Hence, AB cannot be written as a quotient of two integers, which means that AB is an irrational number.

On the other hand, from the Pythagorean theorem it follows that

$$AB^2 = 2,$$

and hence that

$$AB = \sqrt{2}.$$

Therefore, $\sqrt{2}$ *is an irrational number.*

However, the Pythagoreans did not admit that such a number, or a line segment of such length, could exist. The Pythagorean proof used arithmetic, but the situation was described in geometric language by saying that the side and the hypotenuse of a right isosceles triangle are *incommensurable.*

The discovery of irrational quantities was disastrous to the Pythagorean philosophy, according to which the essence of everything could be reduced to natural numbers. There are many legends, some of them conflicting, about this discovery. Some of them deal with the discovery, others deal with the man, sometimes named as Hippasus, who revealed the secret. Revealing Pythagorean discoveries violated the rule that they were to be treated as unwritten secrets of the brotherhood. According to one legend, the anger of the gods led to the death at sea of the person who revealed the secret. Another story says that the discoverer sacrificed an ox in honor of the discovery.

In the beginning the Pythagoreans only found the irrationality of $\sqrt{2}$. We can at once generalize the theorem to: *The square root of any natural number which is itself not the square of a natural number is irrational.* At that time, however, *arithmetica* still consisted largely of the theory of odd and even numbers, and additional tools were needed to prove the generalized theorem. Later (about 400 B.C.), Theodorus proved that $\sqrt{3}, \sqrt{5}, \sqrt{6}, \ldots, \sqrt{15}, \sqrt{17}$, are also irrational. Or, to formulate it in the Greek way, he proved that when (the areas of) two squares are in the ratio of 1:3, or of 1:5, or of 1:6, and so on, then the sides are incommensurable. How Theodorus proved this cannot be stated with certainty. However, from the limited result, which extended no further than $\sqrt{17}$, we may infer that no general method had yet been found, and that for each individual number a proof of its irrationality had to be given. A general method was probably discovered by Theaetetus (about 400 B.C.).

Many methods of proof are known today.

EXERCISES 3-10

1 Prove:
    (a) If a square number is even, it is the square of an even number.
    (b) If a square number is odd, it is the square of an odd number.

2 Prove:
    (a) The sum of two rational numbers is a rational number.
    (b) The product of two rational numbers is a rational number.

3 Prove: If $\sqrt{2}$ is irrational, then $5 + \sqrt{2}$, $5\sqrt{2}$, and $\sqrt{2}/5$ are irrational.

4 Prove:
    (a) The sum of a rational number and an irrational number is an irrational number.
    (b) The product of a rational number and an irrational number is an irrational number.

5 Prove that $\sqrt{3}$ is irrational.

6 Prove that $\sqrt{5}$ is irrational.

7 Read the article cited as reference 25 (this chapter). Several proofs of the irrationality of $\sqrt{2}$ are given.

REFERENCES

For extended discussions of Greek mathematics, see references 4, 7 (Chapter 1), 16 (Chapter 2), and

[18] Dantzig, Tobias, *The Bequest of the Greeks.* New York: Charles Scribner's Sons, 1955.

[19] Heath, T. L., *A Manual of Greek Mathematics.* New York: Oxford University Press, 1931; New York: Dover Publications, Inc., 1963 (paperback).

[20] Heath, T. L., *A History of Greek Mathematics.* New York: Oxford University Press, 1921, 2 vols.

[21] Heath, T. L., *The Thirteen Books of Euclid's Elements*, 2nd ed. New York: Cambridge University Press, 1926, 3 vols.; New York: Dover Publications, Inc., 1956 (paperback). The book contains a translation of the *Elements* with extensive commentary.

Books on the history of mathematics in general, containing chapters on Greek mathematics, are

[22] Boyer, Carl B., *A History of Mathematics*. New York: John Wiley & Sons, Inc., 1968.

[23] Eves, Howard, *An Introduction to the History of Mathematics*, 3rd ed. New York: Holt, Rinehart and Winston, Inc., 1969.

[24] Kline, Morris, *Mathematical Thought from Ancient to Modern Times*. New York: Oxford University Press, 1972.

For some more specialized topics, see

[25] Harris, V. C., "On Proofs of the Irrationality of $\sqrt{2}$," *The Mathematics Teacher*, Vol. 64 (1971), pp. 19-21.

[26] Loomis, Elisha S., *The Pythagorean Proposition*. Washington, D.C.: National Council of Teachers of Mathematics, 1968.

4

THE FAMOUS PROBLEMS
OF GREEK ANTIQUITY

4-1 INTRODUCTION

The three famous problems of the ancient Greeks were to "square" a circle, to trisect an angle, and to duplicate a cube, *by the use of only a straightedge (unmarked) and a compass*. The squaring, or quadrature, of a circle (or another plane figure) means to find a square that has the same area as the region enclosed by the circle (or other figure). These problems were solved, in one sense, by the ancient Greeks themselves. Their solutions required the use of curves other than the straight line and the circle (or mechanical means other than straightedge and compass). However, Greek philosophers, in particular Plato, viewed the straight line and the circle as the basic and perfect curves that should be sufficient to accomplish the constructions. Although, for this reason, the Greeks were dissatisfied with their solutions, these solutions led them to the discovery of a great deal of mathematics. The final "solutions" of the three problems were proofs that the constructions are impossible with only the tools allowed by the philosophers. These proofs were not developed until the nineteenth century and used algebraic ideas that were not known to the Greeks.

4-2   HIPPOCRATES OF CHIOS AND
THE QUADRATURE OF LUNES

Little is known of *Hippocrates of Chios* (about 460-380 B.C.; not to be confused with Hippocrates of Cos, the famous physician). Like Thales, he probably had an early career as a merchant. Aristotle tells us that Hippocrates was swindled by Byzantine customhouse officials, as a result of his simplicity. Others tell us that his vessel was captured by pirates and that he went to Athens to try to recover his property in the law court. History is silent on how his complaint turned out, but he appears to have been in Athens for quite a long time. There he studied geometry and became famous for three major contributions:

1. The quadrature of lunes.
2. An advance toward solution of the problem of duplication of the cube.
3. Compilation of the first textbook of geometry.

Lunes are plane regions bounded by arcs of two different circles, (such as regions I and II in Figure 4-1, the first of the "lunes of Hippocrates"). Hippocrates possibly thought that squaring figures bounded by arcs of circles would lead to a solution of the problem of squaring the circle itself.

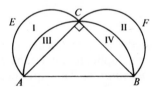

Figure 4-1

At the time of Hippocrates, geometers could construct a square equal in area to the sum of two given squares. They could also construct a square equal in area to a given right triangle and a right triangle equal in area to a given triangle. Since every polygon can be decomposed into triangles, they could combine these constructions to square a polygon (see Exercise 2 of Exercises 4-2).

See Figure 4-1, where the isosceles right triangle ABC is inscribed in the semicircle ACB and where the arcs AEC and BFC are semicircles. Hippocrates is said to have started his work on the quadrature of lunes by proving that the lunes I and II can be squared. He first proved ("$\mathscr{A}$" means "area of") that

$$\mathscr{A}\text{ lune I} + \mathscr{A}\text{ lune II} = \mathscr{A}\,\triangle ABC.$$

He must have reasoned about as follows (we denote the semicircular region bounded by $\overline{AC}$ and arc AEC as s.c.$(\overline{AC})$:

$$\frac{\mathscr{A} \text{ s.c.} (\overline{AC})}{\mathscr{A} \text{ s.c.} (\overline{AB})} = \frac{\pi \left(\frac{AC}{2}\right)^2}{\pi \left(\frac{AB}{2}\right)^2} = \frac{AC^2}{AB^2} = \frac{AC^2}{AC^2 + CB^2}.$$

From this it follows, since $AC = CB$, that

$$\frac{\mathscr{A} \text{ s.c.} (\overline{AC})}{\mathscr{A} \text{ s.c.} (\overline{AB})} = \frac{AC^2}{2AC^2} = \frac{1}{2}$$

and hence that

$$\mathscr{A} \text{ s.c.} (\overline{AC}) = \frac{1}{2} \mathscr{A} \text{ s.c.} (\overline{AB}).$$

Hence, again since $AC = CB$,

$$\mathscr{A} \text{ s.c.} (\overline{CB}) = \frac{1}{2} \mathscr{A} \text{ s.c.} (\overline{AB}),$$

and hence

$$\mathscr{A} \text{ s.c.} (\overline{AC}) + \mathscr{A} \text{ s.c.} (\overline{CB}) = \mathscr{A} \text{ s.c.} (\overline{AB}).$$

From this we find, by subtraction, that

$$
\begin{array}{ccccc}
\mathscr{A} \text{ s.c.} (\overline{AC}) & + & \mathscr{A} \text{ s.c.} (\overline{CB}) & = & \mathscr{A} \text{ s.c.} (\overline{AB}) \\
\hline
\text{area III} & + & \text{area IV} & = & \text{area III} + \text{area IV} \\
\mathscr{A} \text{ lune I} & + & \mathscr{A} \text{ lune II} & = & \mathscr{A} \triangle ABC.
\end{array}
$$

Since lunes I and II are equal in area, each is equal in area to a triangle that is one half of $\triangle ABC$. We know that every triangle can be squared. From this it follows that the lunes in Figure 4-1 can be squared.

EXERCISES 4-2

1 Justify Hippocrates' conclusion by giving the reasons needed to support each step of the derivation above.
2 Carry out each of the following constructions with straightedge and compass.
    (a) Construct a right triangle equal in area to a given oblique triangle. (Hint: Triangles with the same base and altitude have the same area.)
    (b) Construct a rectangle equal in area to a given right triangle.
    (c) Construct a rectangle equal in area to a given triangle.
    (d) Construct a square equal in area to a given rectangle. (Hint: If s is the side of the square and if l and w are the length and the width of the rectangle,

then $s^2 = lw$ and hence $\frac{l}{s} = \frac{s}{w}$. See Figure 4-2, where $ABCD$ is the given rectangle, $CEFG$ is a square, and $\overline{DH}$ is the diameter of a semicircle. Prove that the square is equal in area to the rectangle.)

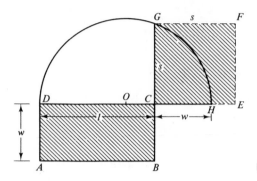

Figure 4-2

(e) Construct a square with area equal to the sum of the areas of two given squares. (Hint: Use the theorem of Pythagoras.)

(f) Construct a square equal in area to a given trapezoid.

(g) Construct a square equal in area to a given pentagon.

(h) Construct a square with area equal to the difference of the areas of two given squares.

(i) Construct a square with area equal to three times the area of a given square. (Hint: See Figure 4-9.)

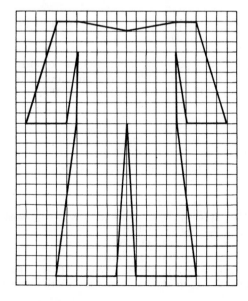

Figure 4-3

3 In medieval Europe, students were given problems that required the geometric construction of squares (or rectangles of a given width) which would have the same area as the pattern for a robe or a cloak. Draw an enlarged copy of the diagram in Figure 4-3, and construct a rectangle with the same area as the pattern and 36 units wide. Then construct a square with the same area as this rectangle.

4 Centuries later, railroad and highway engineers needed a convenient way to calculate the volume of earth that had to be moved from a ditch to be dug or from a cut to be made through the side of a hill. If the cross section of the ditch or the cut would be nearly the same throughout, the volume could be calculated by multiplying the area of the cross section by the length of the excavation. The area of the cross section could be calculated quickly from a surveyor's drawing on coordinate paper (see Figure 4-4), using the formula

The x's and y's represent the coordinates of the vertices of the polygonal cross section taken in a counterclockwise order. The computation follows the lines in the formula, thus:

$$A = \tfrac{1}{2}(x_1 y_2 - x_2 y_1 + x_2 y_3 - x_3 y_2 + \ldots + x_n y_1 - x_1 y_n).$$

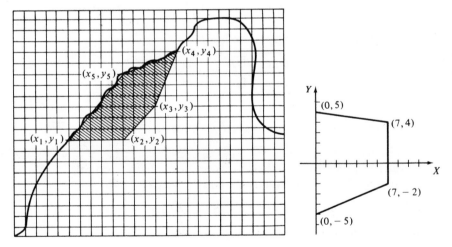

Figure 4-4　　　　　　　Figure 4-5

(a) Apply this method to calculate the area of the trapezoid in Figure 4-5. Check your result by using the formula for the area of a trapezoid.

(b) See Figure 4-3. Choose a pair of axes, assign coordinates to the vertices, and find the area of the polygon.

(c) See (b). In the same way, calculate the shaded area in Figure 4-4.

4-3   OTHER QUADRATURES

Hippocrates next considered a trapezoid $ABCD$ inscribed in a semicircle (see Figure 4-6) such that

$$AD = DC = CB.$$

He also considered a separate semicircle (see Figure 4-7) the diameter $\overline{EF}$ of which is congruent to $\overline{AD}$ in Figure 4-6.

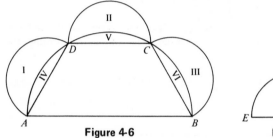

Figure 4-6   **Figure 4-7**

Since $AD = \frac{1}{2}AB$ (why?), we have

$$\frac{\mathscr{A}\,\text{s.c.}\,(\overline{AD})}{\mathscr{A}\,\text{s.c.}\,(\overline{AB})} = \frac{AD^2}{AB^2} = \frac{1}{4}.$$

From this it follows that

$$\mathscr{A}\text{s.c.}(\overline{AD}) + \mathscr{A}\text{s.c.}(\overline{DC}) + \mathscr{A}\text{s.c.}(\overline{CB}) + \mathscr{A}\text{s.c.}(\overline{EF}) = \mathscr{A}\text{s.c.}(\overline{AB}). \quad (1)$$

The circular segments on $\overline{AD}$, $\overline{DC}$, and $\overline{CB}$ marked IV, V, and VI are congruent. If the sum of their areas is subtracted from both members of equation (1), the remainders are equal. Hence,

$$\mathscr{A}\text{lune I} + \mathscr{A}\text{lune II} + \mathscr{A}\text{lune III} + \mathscr{A}\text{s.c.}(\overline{EF}) = \mathscr{A}\text{trapezoid } ABCD. (2)$$

From equation (2) it follows that

$$\mathscr{A}\text{s.c.}(\overline{EF}) = \mathscr{A}\text{trapezoid } ABCD - 3\,\mathscr{A}\text{lune I}. \qquad (3)$$

If Hippocrates could have squared lune I in equation (3), he could have squared the semicircle on $\overline{EF}$ and hence also the complete circle with diameter $\overline{EF}$. Then the problem of squaring the circle would have *been solved.* Starting from the given circle with diameter $\overline{EF}$ and after constructing Figure 4-6, the successive squarings would have been:

1. Lune I (construct a square P such that $\mathscr{A}P = \mathscr{A}$ lune I);
2. A figure that has an area which is equal to 3 times the area of lune I (area equal to $3\mathscr{A}P$);
3. Trapezoid $ABCD$ (construct a square Q such that $\mathscr{A}Q = \mathscr{A}ABCD$);
4. The circle with diameter $\overline{EF}$ (area equal to $2(\mathscr{A}Q - 3\mathscr{A}P)$).

In Section 4-2 we found that some lunes can be squared. In the present section we have seen that the problem of squaring the circle would be solved if lune I in Figure 4-6 could be squared. Unfortunately, this lune is of a different kind than the lunes in Section 4-2. Therefore, Hippocrates had not yet solved the problem of squaring the circle. It is possible that these considerations caused Hippocrates to continue his study of the quadrature of lunes.

In his attempt to square other lunes, Hippocrates used some facts about segments of circles. A segment of a circle is the region bounded by an arc and a chord. Two segments of the same circle or of different circles are said to be similar if the central angles subtended by their chords have equal measure. Hippocrates knew that the areas of similar segments have the same ratio as the squares of the lengths of their chords.

The lunes that we considered so far had a semicircle for their outer perimeter. Hippocrates also considered a lune whose outer perimeter was greater than a semicircle. For this purpose he constructed a special isosceles trapezoid (Figure 4-8) which had the properties that

$$AD = DC = CB$$

and $$AB^2 = 3AD^2.$$

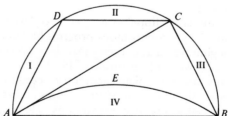

Figure 4-8

He then constructed two arcs. The one through $A, D, C,$ and B is an arc of the circumcircle of $\triangle ABC$. The other, arc AEB, is constructed such that segment IV is similar to segment I. Since the figure is symmetric and $DC < AB$, angles A and B are congruent and acute, with the result that angles C and D are congruent and obtuse. The actual construction of this figure is outlined in Exercise 2 of Exercises 4-4.

Next, Hippocrates proved that the lune bounded by the arcs AEB and ADB is equal in area to the trapezoid $ABCD$ (see Exercise 3 of Exercises 4-4). Therefore, this lune also could be squared.

How did Hippocrates know that the outer circumference of the lune $ADCB$ is greater than a semicircle? He did not merely guess this from Figure 4-8, but he proved it. His proof is as follows:

$$AD^2 + DC^2 < AC^2 \qquad (4)$$

for in $\triangle ACD$, $\angle D$ is obtuse. Further,

$$AB^2 = 3AD^2 = AD^2 + DC^2 + CB^2. \qquad (5)$$

From formulas (4) and (5) it follows that

$$AB^2 < AC^2 + CB^2.$$

Then $\angle ACB$ in $\triangle ABC$ is acute, and therefore arc ACB is greater than a semicircle. In Exercise 1 of Exercises 4-3 the reader is asked to justify the steps in the proof.

Finally, Hippocrates squared a lune whose outer perimeter was smaller than a semicircle. This construction will not be discussed here.

4-4   HIPPOCRATES' GEOMETRY

We have seen that Hippocrates already knew several geometrical theorems. How did he prove them? We do not know. We do know that he tried to prove them, because later writers refer to him as *the first compiler of a geometry textbook;* the Greeks called such a work "Elements of Geometry." We have no detailed data on Hippocrates' book, but we know that it contained much of the contents of the first four books of Euclid's *Elements*.

If we compare Hippocrates' geometry with that of Thales, it appears that in 150 years enormous progress had been made. Thales made a modest beginning: a few theorems, the "proofs" of which (if he did give them) still were of an intuitive kind. Hippocrates showed knowledge of many areas of plane geometry: congruence, similarity, areas, ratios of areas, the theorem of Pythagoras and related theorems, angles in circles, and all kinds of constructions. He may have had but a vague idea of such concepts as ratio and similarity, so that his proofs could not have been flawless, but he made fair progress on the road from intuition to deduction.

Bearing in mind Hippocrates' many merits, we can say that he has been justly classified as one of the great geometers of his time.

EXERCISES 4-4

1 See Figure 4-8. Justify the steps in Hippocrates' proof that arc ACB is greater than a semicircle.

2 See Figure 4-8. In this problem we construct the lune bounded by arcs ACB and AEB.

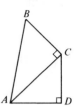

Figure 4-9

(a) See Figure 4-9. Given: $AD = DC = CB$ and $m\angle D = m\angle ACB = 90°$.
Prove: $AB^2 = 3AD^2$.

(b) Suppose that $\overline{AD}$ is a given line segment. Construct a line segment $\overline{AB}$ such that $AB^2 = 3AD^2$.

(c) Use the result obtained in (b) to construct trapezoid $ABCD$ such that $AD = DC = CB$ and $AB^2 = 3AD^2$.

(d) Construct arc ACB and prove that point D lies on this arc.

(e) Construct arc AEB such that segment IV is similar to segment I. (See the definition of similar segments given on page 95.)

3 See Figure 4-8. In this problem we square the lune constructed in Exercise 2.
(a) Prove that the area of segment IV is equal to the sum of the areas of segments I, II, and III.
(b) Prove that the lune and the trapezoid are equal in area.
(c) Square the lune.

4 See Figure 4-8. Prove that diagonal $\overline{AC}$ is tangent to arc AEB.

4-5 DUPLICATION OF THE CUBE

According to one legend, King Minos had built a cube-shaped tomb for his son Glaucus, but when he heard that the tomb was only 100 feet in each direction, he thought this was too small. "It must be doubled in size (volume)," he said, and he ordered the builders to do this quickly by doubling the sides. The mathematicians soon discovered that a mistake had been made, for in this way the new tomb would become eight times the old one. They investigated how it should be done. This turned out to be far from simple.

Another legend connects the problem with the island of Delos, in consequence of which it became known as the *Delian problem*. Apollo was said to have instructed the Delians, by means of an oracle, to double the size of his cube-shaped altar while keeping its form. When they did not succeed in this, they applied to Plato, who told them that Apollo had given this instruction, not because he wanted an altar double the size, but because he wished, in setting this task for them, to point out the importance of mathematics.

Remarkably, the related problem of the duplication of a square does not cause any difficulty. To double a square with side of length a, we must construct a square with side of length x such that

$$x^2 = 2a^2.$$

This equation is equivalent to the proportion

$$\frac{a}{x} = \frac{x}{2a}.$$

Hence, the side of the required square can be constructed as the mean proportional between line segments of lengths a and $2a$.

We can also write

$$x^2 = 2a^2$$

in the form

$$x^2 = a^2 + a^2,$$

from which it follows that x is the length of the diagonal of the given square.

Similarly, to double a cube with edge of length a, we must construct a cube with edge of length x such that

$$x^3 = 2a^3.$$

This problem was reduced by Hippocrates to the problem of constructing two line segments in *continued mean proportion* between segments of length a and $2a$. This means constructing two line segments of length x and y such that

$$\frac{a}{x} = \frac{x}{y} = \frac{y}{2a}.$$

Then the segment of length x is the edge of the desired cube. (See Exercise 2 of Exercises 4-5.)

No one knows what line of thought led Hippocrates to reduce the problem of the duplication of a cube to the construction of two mean proportionals. However, he might have reasoned as follows:

1. Put two cubes with edge of length a against each other such that together they form a block with edges of length $2a$, a, and a (Figure 4-10), which therefore has volume $2a^3$.

2. Imagine this block changed into another block with the same volume and the same height a but such that one of the edges of the base has the required length x (Figure 4-11). Since the volume was to remain the same, the other edge of the base also has to change. Let y represent

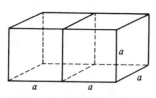

Figure 4-10   **Figure 4-11**

the width of the new base. Since the volume and height of the block do not change, neither does the area of the base. Hence,

$$xy = 2a^2,$$

from which we find that

$$\frac{a}{x} = \frac{y}{2a}. \tag{1}$$

3. Now suppose that the block thus created is changed into a third block, still with the same volume, but with all edges of length x (Figure 4-12). This means that the right face with edges of length a and y is changed into a square with side of length x but with area unchanged. Hence,

$$x^2 = ay,$$

or

$$\frac{a}{x} = \frac{x}{y}. \tag{2}$$

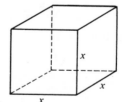

Figure 4-12

4. From formulas (1) and (2) it follows that

$$\frac{a}{x} = \frac{x}{y} = \frac{y}{2a}.$$

The discovery of Hippocrates did not, however, solve the problem of the duplication of the cube. It merely changed the original problem to a new and different one: to construct two mean proportionals between line segments of lengths a and $2a$. We now know that it is impossible to carry out this construction by using straight lines and circles only. In fact, the Greeks, in trying to solve this problem, resorted to the use of other curves or of other instruments besides straightedge and compass. They then found several solutions of the problem. We present two of those solutions.

   a. The solution of Menaechmus   It is known that *Menaechmus* (about 350 B.C.) was the tutor of Alexander the Great. Legend has it that in response to his royal pupil's request for a short cut to geometry, Menaechmus replied, "O, King, for traveling over the country there are royal roads and roads for common citizens; but in geometry there is one road for all."

   Menaechmus is reputed to have discovered the parabola and the equilateral hyperbola. He used them in the solution of the problem of the duplication of the cube. Greek geometers first obtained the three kinds of conic sections by intersecting cones with planes. Later Greeks constructed the curves directly in a plane and named them ellipse, parabola, and hyperbola. Today they are treated as graphs of second-degree equations. Equations were unknown to Menaechmus, but his solution amounts to showing that Hippocrates' problem of finding the mean proportionals could be solved by finding the intersection of two parabolas (or of a parabola and a hyperbola; see Exercise 4 of Exercises 4-5).

   We begin by noting that from Hippocrates' continued mean proportion we can form the two equations

$$\frac{a}{x} = \frac{x}{y} \tag{3}$$

and

$$\frac{x}{y} = \frac{y}{2a}. \tag{4}$$

From equation (3) it follows that

$$x^2 = ay$$

and hence that

$$y = \frac{1}{a}x^2. \tag{5}$$

From equation (4) it follows that

$$y^2 = 2ax$$

and hence that

$$x = \frac{1}{2a}y^2. \tag{6}$$

The graph of equation (5) is a parabola with the origin as its vertex and the y-axis as its axis of symmetry. One of the points of the parabola is the point $P\ (a, a)$. This point can be constructed, for we know the edge of the given cube. Therefore, we know the following data about the parabola, by which it is fully determined: the vertex, the axis of symmetry, and one point.

The graph of equation (6) is also a parabola. Its vertex is also the origin, but its axis of symmetry is the x-axis. One of its points is the point $Q\ (a/2, a)$. Hence, this parabola, also, is determined.

The two parabolas are shown in Figure 4-13.

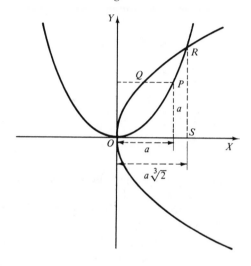

Figure 4-13

The x-coordinates of the points of intersection of the two parabolas are found by solving the equations (5) and (6) simultaneously. Substituting for y from (5) into (6), we have

$$x = \frac{1}{2a}\left(\frac{1}{a}x^2\right)^2,$$

which becomes

$$x^4 - 2a^3x = 0.$$

By factoring the left-hand member we see that this equation has the real roots 0 and $a\sqrt[3]{2}$. The origin, $(0, 0)$, is obviously one point of intersection, but the one we are interested in is $R\ (a\sqrt[3]{2}, a\sqrt[3]{4})$. We drop the perpendicular from R to the x-axis and call the foot S. Now $\overline{OS}$ is of length $a\sqrt[3]{2}$; hence, $\overline{OS}$ is the edge of a cube with volume $2a^3$.

In Menaechmus' construction of the edge of the required cube, we must keep in mind that, although the problem has been solved, the solution was not accomplished by using only straightedge and compass, since parabolas cannot be constructed with these tools.

b. The solution of Plato   The following solution is attributed to *Plato* (about 430-350 B.C.). In Figure 4-14, $\triangle BAD$ and $\triangle ADC$ are right triangles. They have the side $\overline{AD}$ in common, and the hypotenuses $\overline{AC}$ and $\overline{BD}$ intersect at right angles in E. It can be shown (see Exercise 3 of Exercises 4-5) that if $EC = a$, $EB = 2a$, $ED = x$, and $EA = y$, then

$$\frac{a}{x} = \frac{x}{y} = \frac{y}{2a}$$

The reader will recognize the continued mean proportion of Hippocrates.

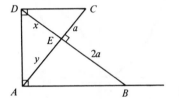

Figure 4-14

Let us now consider the following mechanical device, which will aid in constructing Figure 4-14 if it is required that EC is equal to the length a of given line segment, EB equal to $2a$, and that the angles at A, D, and E are right angles.

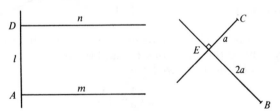

Figure 4-15

In Figure 4-15, all the lines represent rods. Rods m and n are perpendicular to l, and m is rigidly connected to l. The rod n can slide up and down but remains perpendicular to l. The crossed rods at the right are fixed at right angles to each other; $EC = a$ and $EB = 2a$, where a is the length of the edge of the cube to be doubled. The cross must be placed on the system of rods to the left such that C is on n, B is on m, the extension of $\overline{CE}$ passes through A, and the extension of $\overline{BE}$ passes through the intersection D of l and n. By appropriately sliding and rotating the cross and at the same time sliding n, this can always be done. Then $\overline{ED}$ (see Figure 4-14), which is cut off from $\overline{BE}$ extended, is equal to the edge of the required cube. Again we note that the construction is not accomplished by using straightedge and compass only.

Solutions that involve sliding or rotating marked rods until certain conditions are satisfied were called *neusis* or *verging solutions*. Verging solutions required inventiveness and mathematical competency, but they also required the use of tools not allowed under the Platonic restriction to straightedge and compass.

EXERCISES 4-5

1 Given a line segment of length a, construct a line segment of length $a \sqrt{2}$
   (a) by using the construction of a mean proportional,
   (b) by applying the Pythagorean theorem.

2 Given the continued mean proportion $a/x = x/y = y/2a$, prove that $x^3 = 2a^3$.

3 See Figure 4-14.
   (a) Prove: $\triangle DEC \sim \triangle AED \sim \triangle BEA$.
   (b) Prove: $a/x = x/y = y/2a$.

4 (a) Prove that the point of intersection of the parabolas $x^2 = ay$ and $y^2 = 2ax$ (other than $(0, 0)$) also lies on the hyperbola $xy = 2a^2$.
   (b) Graph all three curves on the same axes.

5 Use the method of Menaechmus to find the edge of a cube double in volume to a given cube with an edge of 4 inches.

6 Construct the verging device of Plato (from cardboard or other materials at hand). Use the device to double a cube with an edge of 4 inches. Compare your result with the result obtained in Exercise 5.

4-6   THE TRISECTION PROBLEM

The ancient Greeks were interested in the construction of angles of various measures. Starting with a few fundamental angles such as angles of $60°$ (an angle of an equilateral triangle) and $108°$ (an angle of a regular pentagon), they could construct other specific angles by applications of one or more of the following procedures:

1. Adding two given angles together.
2. Subtracting one given angle from another.
3. Bisecting a given angle.

The Greeks also tried to construct one third of a given angle. They did not succeed in accomplishing this feat with straightedge and compass alone. This is not surprising to us since it was demonstrated (by P. L. Wantzel, in 1837) that there are angles which cannot be trisected. (In the proof, algebraic processes are used by which it was also shown that the duplication of the cube is impossible.)

Several Greek mathematicians invented procedures for trisecting the angle by using verging solutions. We shall describe one by *Archimedes* (287-212 B.C.) and one by *Nicomedes* (about 240 B.C.). A third construction, in which a special curve due to *Hippias* (about 420 B.C.) is used, will be described in the next section.

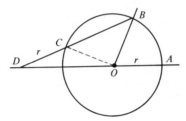

Figure 4-16

a. *The solution of Archimedes*   Let $\angle AOB$ in Figure 4-16 be the given angle. Draw a circle with center O and radius of length r. Extend $\overline{AO}$ in the direction from A to O and draw a line through B in such a way that $CD = r$, where C is on the circle and D on the extension of $\overline{AO}$. Then $m\angle ADB = \frac{1}{3}m\angle AOB$. This is proved as follows. Since

$$DC = CO = OB = r,$$

the triangles DCO and COB are both isosceles. Therefore, $m\angle ODC = m\angle COD$ and $m\angle OCB = m\angle CBO$. Since the measure of the exterior angle of a triangle equals the sum of the measures of the two remote interior angles,

$$
\begin{aligned}
m\angle AOB &= m\angle ODC + m\angle CBO \\
&= m\angle ODC + m\angle OCB \\
&= m\angle ODC + m\angle ODC + m\angle COD \\
&= 3m\angle ODC = 3m\angle ADB.
\end{aligned}
$$

Hence, $$m\angle ADB = \frac{1}{3}m\angle AOB.$$

The verging comes in when the line $\overleftrightarrow{CD}$ is drawn (C on the circle, D on line $\overleftrightarrow{OA}$) so as to satisfy the conditions that it passes through B and that the length of its segment $\overline{CD}$ is r. To perform the construction, mark points D and C on a straightedge such that $DC = r$ and then slide the straightedge in such a manner that D and C remain on the line and circle, respectively, until, finally, line $\overleftrightarrow{DC}$ passes through B.

It is possible to devise instruments that will trisect angles. A sliding linkage that incorporates Archimedes' idea is pictured in Figure 4-17.

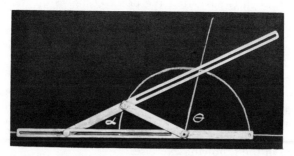

Figure 4-17   Sliding linkage based on Archimedes' trisection process.

   b. The solution of Nicomedes   See Figure 4-18. Angle AOB is the angle to be trisected. Let $OB = a$. Segment $\overline{BC}$ is perpendicular to ray $\overrightarrow{OA}$, and ray $\overrightarrow{BD}$ is parallel to ray $\overrightarrow{OA}$. Ray OPQ has been drawn such that $PQ = 2OB = 2a$. Then $m\angle AOQ = \frac{1}{3}m\angle AOB$.

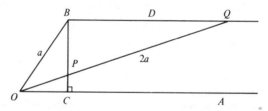

Figure 4-18

   In this case the verging can be done by marking on a straightedge a segment $\overrightarrow{PQ}$ such that $PQ = 2a$ and then sliding P on $\overline{BC}$ and Q on $\overrightarrow{BD}$ until the straightedge passes through O. In Exercise 1 of Exercises 4-6 the reader will be asked to prove that the given angle is trisected.

   Both Archimedes' and Nichomedes' solutions can be accomplished by using a curve called the *conchoid of Nicomedes*. See Figure 4-19. Two points, P and Q, are marked on a straightedge. Let d be the length of $\overline{PQ}$. If the straightedge rotates about the point G and at the same

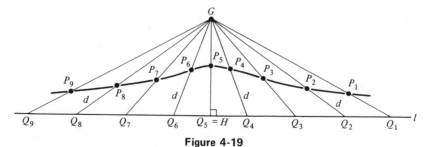

Figure 4-19

time slides through G such that the point Q moves on the line l, then the point P traces a conchoid.

This curve can be used to locate the point P in Figure 4-18. We draw the conchoid with line l of Figure 4-19 along line $\overrightarrow{BD}$, point G at O, and the fixed length d equal to $2a$. Then P is the point of intersection of the conchoid and the perpendicular $\overline{BC}$.

The invention of the conchoid by Nicomedes is an example of something that often happened in Greek mathematics: In trying to solve a geometrical problem, the Greeks were led to invent and study an interesting new curve. Further uses of and facts about the conchoid are suggested in the following exercises.

EXERCISES 4-6

1 See Figure 4-18. Prove that $mLAOQ = \frac{1}{3}mLAOB$. (Hint: Connect B with the midpoint of $\overline{PQ}$.)

2 Construct an angle of $60°$ and trisect it by use of Archimedes' verging construction. Use a protractor to check the accuracy of your work.

3 Repeat Exercise 2; use Nicomedes' verging construction.

4 Draw the conchoid of Figure 4-19. Use $d = 2$ (inches) and $GH = \frac{1}{2}\sqrt{3}$ (inches).

5 Use the conchoid drawn in Exercise 4 to trisect an angle of $60°$
   (a) by drawing the angle and the conchoid in proper relation to each other on the same paper,
   (b) by drawing the angle on transparent paper and superimposing the angle on the conchoid drawn in Exercise 4.

6 (a) Explain the specification $d = 2$, $GH = \frac{1}{2}\sqrt{3}$ in Exercise 4, if the conchoid is to be used for trisecting an angle of $60°$. (Hint: $\frac{1}{2}\sqrt{3} = \sin 60°$.)
   (b) Can the same conchoid be used to trisect other angles?

7 Today the conchoid is defined differently. It has another branch in addition to the one shown in Figure 4-19. The other branch is obtained by locating, for each value of i, the point R_i on the ray $\overrightarrow{GQ_i}$ such that $Q_iR_i = d$ and R_i is on the other side of Q_i from P_i. Trace the figure you drew in Exercise 4 and construct its other branch.

In Exercises 8-12, "conchoid" will mean the conchoid as defined in Exercise 7.

8 Given: a line l and a point G such that $GH = 2$, where H is the foot of the perpendicular from G to l. Draw (in the same figure, using different colors for each case) conchoids for each of the following values of d:
   (a) 1 (inch)
   (b) 2 (inches)
   (c) 4 (inches)

9 See Figure 4-19. Show that the polar equation of a conchoid is: $\rho = a \sec \theta \pm d$, where $\rho = GP$, $a = GH$, and $\theta = mLHGP$. (Hint: Choose $\overrightarrow{GH}$ as the positive x-axis.)

10   See Exercise 9. Graph conchoids for the following values of a and d:
(a) $a = 1, d = \frac{1}{2}$
(b) $a = 1, d = 1$
(c) $a = 1, d = 2$

11   The equation of a conchoid in Cartesian coordinates is $(x - a)^2(x^2 + y^2) = d^2 x^2$. Derive this equation by use of
(a)   the definition of the conchoid,
(b)   the equation of the conchoid in polar coordinates.

12   See the conchoids that you drew in Exercise 10.
(a)   What kind of symmetry do they have?
(b)   Use the equation obtained in Exercise 9 to prove your answer in (a).
(c)   Use the equation obtained in Exercise 11 to prove your answer in (a).

4-7   HIPPIAS AND SQUARING OF THE CIRCLE

The duplication and trisection problems are closely related:

1. They can both be solved by using conic sections (see Section 4-5, the solution of Menaechmus, and Exercise 16 of Exercises 4-7).
2. When expressed algebraically, they both lead to cubic equations (see Section 4-8).
3. The proofs that they cannot be solved with straightedge and compass alone make use of the same approach and were given at the same time (see Section 4-8).

The squaring of the circle is also impossible with straightedge and compass alone. The nature of the proof that this is impossible is quite different, however, from the nature of the proofs referred to in point 3, and it did not evolve until 1882. However, an ancient Greek solved both the problem of squaring the circle and the problem of trisecting the angle with the same special curve. The curve was invented by the sophist *Hippias* (about 420 B.C.) and is called the *quadratrix*. It is constructed as follows (see Figure 4-20). We start from a square $ABCD$. A line segment, which first coincides with $\overline{AB}$, turns about A from position $\overline{AB}$

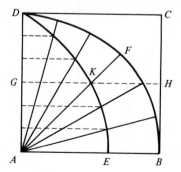

Figure 4-20

to position $\overline{AD}$. A second line segment, also coinciding first with $\overline{AB}$, moves vertically upward, remaining parallel to $\overline{AB}$, to position $\overline{DC}$. The two line segments start moving at the same moment. The first rotates at a constant angular velocity. The second slides upward at a constant linear velocity. The first coincides with $\overline{AD}$ at the same moment that the second coincides with $\overline{DC}$. At each moment during their simultaneous movement the two line segments have a point of intersection. *The set of all these points is the quadratrix.* For example, when the rotating line segment reaches position $\overline{AF}$ such that $m\angle BAF = 45°$, or half the rotation from $\overline{AB}$ to $\overline{AD}$, the sliding line segment will reach position $\overline{GH}$, where G is the midpoint of $\overline{AD}$. Their point of intersection, K, is then a point of the quadratrix.

We can construct additional points of the quadratrix as follows. Divide $\angle DAB$ into six congruent parts by drawing radii making angles of $30°$, $45°$, and $60°$ with $\overline{AB}$ and then using these angles to determine radii making angles of $15°$ and $75°$ with $\overline{AB}$. Construct points on $\overline{AD}$ that divide it into six equal parts. Draw lines through these points, parallel to $\overline{AB}$. The intersections of these lines with the corresponding radii are points of the quadratrix. We can find any number of additional points of the quadratrix by bisecting corresponding angles and segments and finding the point of intersection of each radius and the corresponding parallel line.

From the preceding discussion we see that *the defining property of the quadratrix is as follows* (see Figure 4-21):

$$\frac{m\angle XAB}{m\angle DAB} = \frac{XX'}{DA} \tag{1}$$

We proceed to show that a given angle can be trisected and a given circle squared with the use of a quadratrix.

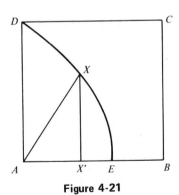

Figure 4-21

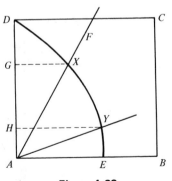

Figure 4-22

    a. Trisection of the angle using the quadratrix   See Figure 4-22, where the given angle BAF is trisected as follows:

1. Construct a square $ABCD$ on $\overline{AB}$, such that ray $\overrightarrow{AF}$ lies in the interior of $\angle BAD$.
2. Draw the quadratrix DE and let X be its point of intersection with $\overrightarrow{AF}$.
3. Draw $\overline{XG}$ parallel to $\overline{BA}$.
4. Construct the point H on $\overline{AD}$ such that $AH = \frac{1}{3}AG$.
5. Draw $\overline{HY}$ parallel to $\overline{AB}$ (Y on the quadratrix).
6. Draw $\overrightarrow{AY}$.

Then $\angle BAY$ is the required angle (see Exercise 1 of Exercises 4-7).

    Of course, we must keep in mind that the quadratrix as a whole cannot be constructed with straightedge and compass and that therefore the angle-trisection problem has not been solved.

    b. Quadrature of the circle using the quadratrix   Pappus (about A.D. 320) reports that Dinostratus (about 350 B.C.), the brother of Menaechmus, used the quadratrix as a means to square the circle. It is possible that Hippias himself did not know that his curve could be used for that purpose.

    Pappus, and probably Dinostratus himself, proves (see Figure 4-20) that

$$\frac{l\,(\text{arc } BFD)}{AB} = \frac{AB}{AE}, \tag{2}$$

where $l(\text{arc } BFD)$ means the length of arc BFD.

    We shall simplify our notation by indicating

$$AB \text{ by } r$$

(since $\overline{AB}$ is a radius of the circle of which BFD is an arc),

$$l(\text{arc } BFD) \text{ by } q$$

(since it is the length of a quarter circle), and

$$AE \text{ by } e.$$

Then proportion (2) takes on the form

$$\frac{q}{r} = \frac{r}{e}. \tag{3}$$

    We postpone the proof of proportion (3) and use this proportion to see how we can square the circle with radius of length r if we know the place of point E. As a first step toward squaring the circle, it is convenient to construct a line segment equal in length to the circumference of the circle. (This process is called *rectification of the circle*.) From proportion (3), which can be written in the form

$$\frac{e}{r} = \frac{r}{q},$$

it follows that q, the length of the quarter circle, is the fourth proportional to e, r, and r. Therefore, we can construct a line segment of length q (see Exercise 7 of Exercises 4-7). The circle is then rectified by constructing a line segment of length $4q$. That is, $C = 4q$, where C is the circumference of the circle.

For the purpose of squaring the circle, we shall use the following theorem, stated by Archimedes but almost certainly known to Dinostratus: *The area of a circle is the same as the area of a right triangle that has an altitude with the same length as a radius and a base with a length equal to the circumference of the circle.*

We now construct a right triangle with base of length C and altitude of length r, and we convert this triangle to a square (see Exercise 2(d) of Exercises 4-2). Then the circle has been squared.

We still have to show how Pappus proved proportion (3). He used an indirect proof. He showed that q/r can neither be less than nor greater than r/e. We begin by proving the first part.

Assume that

$$\frac{q}{r} < \frac{r}{e}. \tag{4}$$

Then there exists a number, a, such that $a > e$ and that

$$\frac{q}{r} = \frac{r}{a}. \tag{5}$$

Since $q > r$, it follows from proportion (5) that

$$r > a.$$

Hence, we have

$$r > a > e.$$

This means (see Figure 4-23) that there is a segment, $\overline{AP}$, of length a such that

$$AB > AP > AE.$$

Therefore, the point P is between B and E.

We draw the quarter-circle with center A and radius a. Let Q and

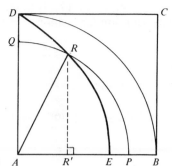

Figure 4-23

R be its points of intersection with $\overline{AD}$ and the quadratrix DE, respectively. Then the lengths of the quarter-circles BD and PQ are proportional to the lengths of the radii $\overline{AB}$ and $\overline{AP}$. Hence,

$$\frac{q}{r} = \frac{l\,(\text{arc } PQ)}{a}.$$

From this proportion and proportion (5) it follows that

$$l\,(\text{arc } PQ) = r. \tag{6}$$

Drop the perpendicular $\overline{RR'}$ from R to $\overline{AB}$. Since R is a point of the quadratrix, it satisfies the quadratrix condition (equation (1)). Therefore, we have

$$\frac{m\angle BAR}{m\angle BAD} = \frac{RR'}{DA},$$

and hence

$$\frac{l\,(\text{arc } PR)}{l\,(\text{arc } PQ)} = \frac{RR'}{DA}.$$

It then follows from equation (6) and the fact that $DA = r$ that

$$\frac{l\,(\text{arc } PR)}{r} = \frac{RR'}{r},$$

from which we conclude that

$$l\,(\text{arc } PR) = RR'. \tag{7}$$

However, this is impossible since RR', the length of the perpendicular from R to $\overline{AB}$, is the shortest distance from R to $\overline{AB}$. Hence, the assumption in statement (4),

$$\frac{q}{r} < \frac{r}{e},$$

from which we derived equation (7), is false.

Pappus proved in a similar fashion that the inequality

$$\frac{q}{r} > \frac{r}{e} \tag{8}$$

is also false. (We leave the proof to the reader; see Exercise 13 of Exercises 4-7).

Since neither of the assumptions (4) and (8) is true, it follows that

$$\frac{q}{r} = \frac{r}{e},$$

which proves proportion (3).

Recall that we have shown, following (3), that if (3) is true, then the circle can be squared. However, this construction uses segment $\overline{AE}$. But it turns out to be impossible to find point E with straightedge and compass. E can only be approximated by repeatedly bisecting angles with side $\overrightarrow{AB}$ and the corresponding segments of $\overline{AD}$ in order to construct points of the quadratrix closer and closer to E. Therefore, neither has the given circle been squared with straightedge and compass only. Hence, the problem of squaring the circle has not been solved.

EXERCISES 4-7

1 See Figure 4-22. Given: $AH = \frac{1}{3}AG$. Prove: $m\angle BAY = \frac{1}{3}m\angle BAX$.

2 Draw a quadratrix in a square with sides of 3 inches. (Make an accurate drawing after constructing at least 15 points. This quadratrix will be used as an underlay in subsequent exercises.)

3 (a) Construct an angle of $60°$.
   (b) Trisect this angle by using the quadratrix of Exercise 2.

4 Repeat Exercise 3 for an angle of $75°$.

5 Draw an acute angle and trisect it by using the quadratrix of Exercise 2.

6 Draw an acute angle and divide it into five equal parts by using the quadratrix of Exercise 2.

7 Figure 4-24 shows the construction of the fourth proportional (with length x) to the given line segments with lengths a, b, and c. (In $a/b = c/x$, x is the fourth proportional to a, b, and c.) List the steps by which point E was determined. Prove that the construction does produce a line segment with a length that satisfies the proportion.

8 Assume a unit length. By using a quadratrix, construct a square equal in area to a circle with radius of length 2. Measure the side of the resulting square. Check the outcome by using the formulas for the area of a square and of a circle.

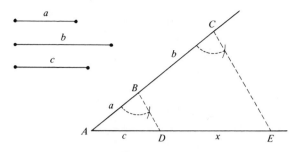

Figure 4-24

9 (a) The astronomer *Johannes Kepler* (1571-1630) explained the formula for
the area of a circle by use of a drawing similar to the one in Figure 4-25 (a).
Assume that the sectors represent n small congruent isosceles triangles;
find the area of the circle in terms of the radius r and the perimeter p by
adding the areas of the triangles together. (Hint: Assume that the altitude
of each of the triangles is equal to r.)

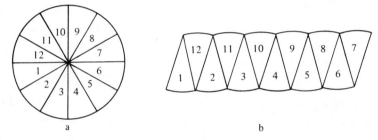

Figure 4-25

(b) Schoolchildren are sometimes introduced to the concept of the area of a
circle through a procedure in which the sectors in Figure 4-25 (a) are cut
out and rearranged as in Figure 4-25 (b). Assume that Figure 4-25 (b) represents a parallelogram; calculate its area in terms of r and p. (Hint:
Assume that the altitude of the parallelogram has length r.)

(c) How do the outcomes in (a) and (b) lead to Archimedes' theorem (see
page 110).

(d) Assume that the area of a circle is πr^2 and verify Archimedes' theorem
(see page 110).

10 See Figure 4-26. Show that the equation of the quadratrix in polar coordinates
with the pole at A and $\overrightarrow{AB}$ as the polar axis is

$$\rho = \frac{2r\theta}{\pi \sin \theta}.$$

11 See the equation in Exercise 10. Compare the value of ρ for $\theta = \pi/2$ with
(a) the value of XX' in equation (1) (page 108) when $m\angle XAB = 90°$,
(b) the y-coordinate of D read from Figure 4-26.

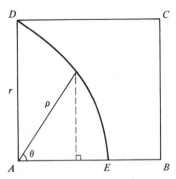

Figure 4-26

12 See the equation in Exercise 10. Show that
    (a) the value of ρ is undefined for $\theta = 0$,
    (b) $\lim\limits_{\theta \to 0} \rho$ is equal to $2r/\pi$,
    (c) $r = \frac{1}{2}\pi e$, where $e = AE$.

13 See Dinostratus' proof that the statement $q/r < r/e$ is false. Prove, by use of similar geometric arguments, that the statement $q/r > r/e$ is false.

14 Assume a unit length of 1 inch.
    (a) Construct a line segment of length π by use of the quadratrix in a square of side 1.
    (b) Construct a line segment of length π by use of the quadratrix of Exercise 2.

15 There is another Greek method for trisecting an angle. It uses a hyperbola. See Figure 4-27, which we build up as follows. (Draw an accurate figure on squared paper. It will be used as an underlay in Exercise 17.)
    (a) Draw a vertical line, d.
    (b) Select a point, F, at a distance of about 1 inch from d.
    (c) Drop the perpendicular $\overline{FQ_0}$ from F to d (Q_0 on d) and construct on $\overline{FQ_0}$ a point, P_0, such that $FQ_0 = 2P_0Q_0$.
    (d) Construct several points, P_i such that for each value of i, $FP_i = 2P_iQ_i$, where P_iQ_i is the distance from P_i to d. Then we have, for each value of i,

$$\frac{FP_i}{P_iQ_i} = 2. \tag{9}$$

    (e) Draw a smooth curve through the points P_i. This curve is one of the two branches of a hyperbola. (Since we shall not need the other branch, we shall not construct it.) The ratio in formula (9) is called the *eccentricity* of the curve, d the *directrix*, and F the *focus*. (The curve is a hyperbola, a parabola, or an ellipse depending on whether the eccentricity is greater than, equal to, or less than 1. For trisecting the angle we shall only need the curve with eccentricity 2.)

16 See Figure 4-28. Let the curve PQ be a branch of the hyperbola with focus F, directrix $\overleftrightarrow{BD}$ and eccentricity 2. Let ABC be an angle such that its side $\overrightarrow{BA}$

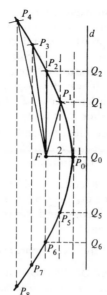

Figure 4-27

passes through F and its bisector coincides with $\overrightarrow{BD}$. The circle with center B and radius $\overline{BF}$ intersects the hyperbola in a point P. Prove that $m\angle ABP = \frac{1}{3}m\angle ABC$ in two ways, (a) and (b).

(a) Use the following three line segments: $\overline{PF}$, the perpendicular from B to $\overline{PF}$, and the perpendicular from P to $\overrightarrow{BD}$.

(b) Draw the other trisecting ray, $\overrightarrow{BP'}$, of $\angle ABC$ (P' on the circle), and use the hyperbola with focus G, directrix $\overleftrightarrow{BD}$, and passing through P'. (Hint: Consider the chords $\overline{FP}$, $\overline{PP'}$, and $\overline{P'G}$ of the circle.)

17 Draw an arbitrary angle and trisect it by using the hyperbola of Exercise 15.

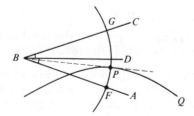

Figure 4-28

4-8 THE SOLUTIONS OF THE GREEK PROBLEMS

The duplication, trisection, and quadrature problems were "solved" in the nineteenth century. In each case, the "solution" turned out to be a proof that the construction called for was impossible. The proofs of the impossibility of duplicating the cube and trisecting the angle by using straightedge and compass only will be outlined in the remaining part of this chapter. These proofs depend on algebraic ideas that were developed over many centuries, in particular on the theory of cubic equations. The proof that the quadrature of the circle is impossible is more difficult and will be omitted. It was developed later than the other two proofs, when the German mathematician *F. Lindemann* (1852-1939) showed, in 1882, that the number π is transcendental.

To show that duplication and trisection are impossible, we will need two theorems, the *constructible-root theorem and the rational-root theorem*. We first give the following:

DEFINITION *By a* constructible root *of an equation we mean a root that has the following property: If a unit of length is given, we can construct with straightedge and compass a line segment the length of which is equal to the root.*

Then the first of the two theorems is the following:

CONSTRUCTIBLE-ROOT THEOREM *A cubic equation with integral coefficients that has no rational root has no constructible root.*

The proof is beyond this book. However, as we shall see, it is easy to apply the theorem itself.

The second theorem concerns polynomial equations. Examples of these are:

$$2x + 3 = 0 \qquad (1)$$
$$x^2 - 2x - 3 = 0 \qquad (2)$$
$$2x^3 - x^2 - 7x + 6 = 0 \qquad (3)$$
$$3x^4 - 10x^3 + 2x - 3 = 0. \qquad (4)$$

The general form of a polynomial equation is

$$a_0 x^n + a_1 x^{n-1} + a_2 x^{n-2} + \cdots + a_{n-1} x + a_n = 0,$$

in which $a_0, a_1, a_2, \ldots, a_n$ are the coefficients.

The second theorem is the following:

RATIONAL-ROOT THEOREM *If the coefficients of a polynomial equation are integers, then any rational root of the equation can be written in the form p/q, where p is a factor of a_n and q is a factor of a_0.*

For example, in equation (1), $a_n = 3$ and the factors of 3 are ± 1 and ± 3. In the same equation, $a_0 = 2$ and the factors of 2 are ± 1 and ± 2. Therefore, by the rational-root theorem, the only possible rational roots of (1) are

$$\frac{\pm 1}{\pm 1} = \pm 1, \qquad \frac{\pm 1}{\pm 2} = \pm \frac{1}{2}, \qquad \frac{\pm 3}{\pm 1} = \pm 3, \qquad \frac{\pm 3}{\pm 2} = \pm \frac{3}{2}.$$

It is easy to see that the only root of equation (1) is $-\frac{3}{2}$.

In equation (2), the factors of -3 are ± 1 and ± 3. Those of 1, the coefficient of x^2, are ± 1. Hence, the possible rational roots of (2) are

$$\pm 1 \quad \text{and} \quad \pm 3.$$

Show, by substituting these values in equation (2), that 3 and -1 are roots whereas -3 and 1 are not.

Of course, we could easily have solved equations (1) and (2) directly, without using the rational-root theorem. However, solving equation (3) directly is much more difficult than by using this theorem. The possible rational roots of (3), as formed from the factors of 6 and 2, are

$$\pm 1, \quad \pm \frac{1}{2}, \quad \pm 2, \quad \pm 3, \quad \pm \frac{3}{2}, \quad \pm 6.$$

Show that $1, -2$, and $\frac{3}{2}$ are roots and that none of the other possibilities satisfies the equation.

The rational-root theorem does not help us in finding irrational roots. For example, the roots of

$$x^2 - 2x - 1 = 0 \tag{5}$$

are $$1 + \sqrt{2} \quad \text{and} \quad 1 - \sqrt{2},$$

both of which are irrational. The theorem tells us only that the *possible* rational roots are ± 1. Substitution of 1 and -1 for x in equation (5) shows that neither of them is a root and that therefore this equation has no *rational* roots.

We shall omit a proof of the rational-root theorem.

We can now show that duplication and trisection are impossible.

a. Duplication　Choose a unit of length equal to the edge of the given cube. Then the volume of the cube is $1 \ (= 1^3)$. Therefore, the required cube has the volume 2. We want to construct a line segment that has the same length as the edge of this cube. If we represent this length by x, we must have

$$x^3 = 2. \tag{1}$$

To show that duplication of our cube is impossible, we will show that equation (1) has no constructible root.

Equation (1) is equivalent to

$$x^3 - 2 = 0. \tag{2}$$

By the rational-root theorem, the possible rational roots of this equation are ± 1 and ± 2. Since none of these values satisfies equation (2), this equation has no *rational* roots. Therefore, (2) is a cubic equation with integral coefficients that has no rational roots, and we can apply the constructible-root theorem. Hence, we conclude that (2) has no constructible root and that the cube cannot be duplicated.

b. Trisection   To prove that the trisection of the angle is impossible, we only have to show one angle that cannot be trisected. For that purpose we recall Nicomedes' method for trisecting an angle (see Figure 4-18). Figure 4-29 is obtained from Figure 4-18 by constructing the line segments $\overline{BR}$ (R the midpoint of $\overline{PQ}$) and $\overline{BE}$ (perpendicular to line $\overleftrightarrow{OQ}$).

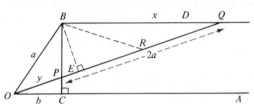

Figure 4-29

We have from the previous discussion (see page 105):

$$OB = a$$
$$PQ = 2a,$$

from which it follows that

$$BR = PR = QR = a.$$

We further indicate

$$OC \quad \text{by} \quad b$$
$$BQ \quad \text{by} \quad x$$
$$OP \quad \text{by} \quad y.$$

Now we can easily prove that

$$\triangle PBQ \sim \triangle PCO \sim \triangle BEQ,$$

from which we obtain the following continued proportion:

$$\frac{BQ}{PQ} = \frac{CO}{PO} = \frac{EQ}{BQ}. \tag{3}$$

Since $\triangle ORB$ is isosceles, with $BO = BR$, the foot of the perpendicular $\overline{BE}$ on $\overline{OR}$ is the midpoint of $\overline{OR}$. Therefore,

$$ER = \frac{1}{2}OR$$
$$= \frac{y + a}{2}$$

and
$$EQ = ER + RQ$$
$$= \frac{y + a}{2} + a$$
$$= \frac{y + 3a}{2}.$$

Hence, (3) takes the form

$$\frac{x}{2a} = \frac{b}{y} = \frac{y + 3a}{2x} \tag{4}$$

In Figure 4-29 we chose point B on one side of the given angle and denoted OB by a. Suppose, for convenience, that we take $a = 1$. Then equations (4) take the form

$$\frac{x}{2} = \frac{b}{y} = \frac{y + 3}{2x}. \tag{5}$$

By equating the first expression in (5) to each of the other two, we get the following equations:

$$xy = 2b$$
$$x^2 = y + 3.$$

Substituting for y from the first equation into the second, we obtain the equation

$$x^3 - 3x - 2b = 0. \tag{6}$$

We could call equation (3) a "trisection equation." For if it has a constructible root, x_1, we can construct a line segment with length x_1. This would make it possible to locate the point Q and hence the trisector $\overrightarrow{OQ}$ of $\angle AOB$.

Whether or not equation (6) has a constructible root depends upon the value of b. For instance, if $b = 0$, equation (6) becomes

$$x^3 - 3x = 0,$$

or $$x(x^2 - 3) = 0,$$

which has the roots 0, $\sqrt{3}$, and $-\sqrt{3}$. All these roots are constructible. In particular, the root $\sqrt{3}$ is constructible (see Exercise 5 of Exercises 4-8). This means that if we locate Q on $\overrightarrow{BD}$ such that $BQ = \sqrt{3}$, then $\overrightarrow{OQ}$ trisects $\angle AOB$. (Of course, $\angle AOB$ could easily have been trisected without constructing $\overline{BQ}$. If $b = 0$, points O and C coincide, and $m\angle BOC = 90°$. One third of $90°$ is $30°$, and an angle of $30°$ can be constructed as one half of an angle of an equilateral triangle. Hence, an angle of $90°$ can be trisected.)

There are, in fact, infinitely many angles that can be trisected. Every angle that we can construct is one third of some other angle, and therefore that other angle can be trisected. (For example, since we can construct an angle of 30°, we can construct an angle of 15° (by bisection), and therefore an angle of 45° can be trisected.)

Let us now consider the case $b = \frac{1}{2}$ in equation (6). Then the equation takes the form

$$x^3 - 3x - 1 = 0. \tag{7}$$

Its possible rational roots are ±1. Since neither of these values satisfies (7), this equation has no rational roots. Hence, by the constructible-root theorem, it has no constructible roots, and we cannot construct a line segment $\overline{BQ}$ that satisfies the requirement. We therefore cannot trisect an angle for which $b = \frac{1}{2}$. But (in Figure 4-29)

$$b = \frac{b}{1} = \frac{OC}{OB} = \cos \angle AOB.$$

Therefore, if $b = \frac{1}{2}$, we have $m\angle AOB = 60°$. Hence, we have shown that an angle of 60° cannot be trisected.

Since we found an angle that cannot be trisected, we have proved that the trisection of the angle with straightedge and compass only is impossible.

Thus, a long trail, beginning with ancient Greek geometric problems, ended in modern algebraic concepts. Mathematicians do not feel disappointed that these constructions turned out to be impossible. They are pleased that three interesting problems are solved and that along the more than 2000-year-long path many unforeseen and interesting developments have grown up.

EXERCISES 4-8,

1 Find the rational roots, if any, of the following equations:
   (a) $5x - 2 = 0$ (b) $5x^2 - 2 = 0$
   (c) $5x^2 - 125 = 0$ (d) $x^2 - 4 = 0$
   (e) $4x^2 - 25 = 0$ (f) $x^2 - 5x + 6 = 0$
   (g) $x^2 + 5x - 6 = 0$ (h) $x^2 - 5x - 6 = 0$
   (i) $x^2 + 5x + 6 = 0$ (j) $2x^2 - x - 3 = 0$
   (k) $x^3 - 6x^2 + 11x - 6 = 0$ (l) $x^3 - 4x^2 + x + 6 = 0$
   (m) $2x^3 - 5x^2 - x + 6 = 0$

2 One root of the equation $x^n - a = 0$, where $a > 0$, is $\sqrt[n]{a}$.
   (a) Use the equation $x^2 - 5 = 0$ to show that $\sqrt{5}$ is irrational.

(b) Show that $\sqrt[3]{5}$ is irrational.

(c) Show that $\sqrt[n]{5}$ is irrational for $n > 2$.

(d) Show that $\sqrt{p}$ is irrational if p is a prime number.

(e) Show that $\sqrt[n]{p}$ is irrational if p is a prime number and $n > 2$.

3 Let line segments of lengths 1, a, and b be given. Construct a line segment of length ab. (Hint: See Figure 4-30.)

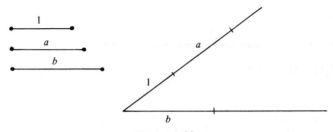

Figure 4-30

4 Construct a line segment of length a/b; use the same data as in Exercise 3.

5 Construct a line segment of length $\sqrt{a}$ by using the line segments of lengths 1 and a given in Exercise 3. (Hint: In Figure 4-31, $AD \cdot BD = CD^2$. Now let $AD = a$ and $BD = 1$.)

6 See Figure 4-32.

(a) Construct a line segment of length $\sqrt{17}$.

(b) Prove that for any positive integer n, we can construct a line segment of length $\sqrt{n}$.

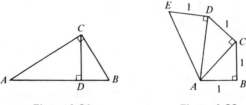

Figure 4-31   **Figure 4-32**

REFERENCES

For discussions of the famous problems of Greek mathematics, see references 7 (Chapter 1), 19 and 20 (Chapter 3), and (especially for proofs of the impossibility of their solution)

[27] Courant, Richard, and Herbert Robbins, *What Is Mathematics?* New York: Oxford University Press, 1947.

[28] Yates, Robert C., *The Trisection Problem.* Washington, D.C.: National Council of Teachers of Mathematics, 1971.

5

EUCLID'S PHILOSOPHICAL FORERUNNERS

5-1 PHILOSOPHY AND PHILOSOPHERS

Pythagoras and his followers were philosophers, "lovers of ideas." As we have seen, they constructed a semimystical system of ideas that explained the physical world and the universe in terms of whole numbers. Their philosophy also led them to be vegetarians and to believe in the transmigration of souls.

Philosophy is related to all branches of knowledge, since it is concerned with ultimate causes and with values. Although most philosophers do not do mathematics, many of them have greatly influenced the growth of mathematics. Philosophers ask: What are the typical features of a subject, such as mathematics, and what are the general characteristics of mathematical ways of thinking? Mathematicians formulate definitions and axioms and prove theorems. Mathematicians may be set to thinking by observing the world around them and its real-life problems. Some of their theories can be interpreted and applied in the physical world. However, most mathematicians accept the modern philosophical ideas that their axioms are logically arbitrary and that their theorems are about mental concepts. These mental concepts cannot be actually observed in the physical world. This view of the nature of mathematics can be traced back to the Greek philosopher Plato.

Of course, the same person can be both a philosopher and a mathematician. In recent years, *Bertrand Russell* and *Alfred North Whitehead* were both. *René Descartes* (1596-1650) and *Gottfried Wilhelm Leibniz* (1646-1716) were prominent in both fields. The most famous Greek philosophers were *Socrates* (469?-399 B.C.), *Plato* (427-347 B.C.) and *Aristotle* (384-322 B.C.). The work of none of them was as mathematical as that of Pythagoras, but Plato and Aristotle had important effects upon the growth of mathematics. Note that their lives overlapped. Socrates did not write his philosophy down, but Plato, who did write extensively, preserved and explained many of Socrates' ideas. Aristotle did not know Socrates.

5-2 PLATO

Plato wrote much of his philosophy in the form of reports of conversations, or *dialogues*, between his old teacher Socrates and various other persons, such as Phaedo and Meno. At one point, Plato had Socrates, while talking to Meno, call over a slave boy who had never studied mathematics. Plato records as follows the beginning of the dialogue in which Socrates led the boy first to conclude, falsely, that to double the area of a square one would double the sides, and then to see his error and correct it. We have used "feet" and inserted a few words to make the dialogue easier to follow for modern readers.

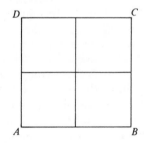

Figure 5-1

Socrates: Tell me, boy, do you know that a figure like this (Figure 5-1) is a square?

Boy: I do.

Socrates: And do you know that a square figure has these four segments equal?

Boy: Certainly.

Socrates: And these segments which I have drawn through the middle of the square are also equal?

Boy: Yes.

Socrates: A square may be of any size?

Boy: Certainly.

Socrates: If in one direction the space was of 2 feet and in the other direction of 1 foot, the whole would be of 2 square feet taken once?

Boy: Yes.

Socrates: But since this side is also of 2 feet, there are twice 2 square feet?
Boy: There are.
Socrates: Then the area of the square is of twice 2 square feet?
Boy: Yes.
Socrates: And how many are twice 2 square feet? Count and tell me.
Boy: Four, Socrates.
Socrates: And might there not be another square twice as large in area as this, and having like this the lines equal?
Boy: Yes.
Socrates: And of how many square feet will that be?
Boy: Of 8 feet.
Socrates: And now try and tell me the length of the line that forms the side of that double square: this is 2 feet—what will that be?
Boy: Clearly, Socrates, that will be double.

At this point Socrates had led the slave boy into the error that the side of the square with an area of 8 square feet will be 4 feet, twice the length of the side of a square with an area of 4 square feet. By computation he next led the slave boy to discover that when the side was doubled, the area became not twice but four times as large. The slave then proposed to make the new side one and one half times as long. Again the result was wrong. Socrates paused to emphasize to his friend Meno that the slave had already obtained much more insight into the problem than he showed at the beginning of the dialogue, although no knowledge had been imparted to him. The slave's added insight was developed by the questions put to him. Socrates thought that these questions had made the slave aware of his latent knowledge. In Socrates' view the slave knew all these things, but he was not aware that he knew them. As a result of the questioning, this knowledge had returned.

Socrates finally led the slave to see the solution of the problem. He drew four squares with sides 2 feet long, which together formed one square with sides of 4 feet. In these squares (see Figure 5-2), he drew four diagonals. They enclosed the required square, a square with twice the area of the original square.

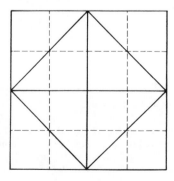

Figure 5-2

Plato's real purpose in writing this dialogue was not to develop a particular piece of mathematics. His purpose was to develop one of the beliefs of his philosophy, that teaching is only aimed at reawakening in the mind of the student the memory of something previously known. According to this belief, Socrates did not teach the slave boy anything new. Socrates' questions led the boy to recall ideas that had been in his mind already before birth. This suggests another of Plato's philosophical beliefs. It is the belief that, although material objects or problems may start us thinking and that diagrams may help, the ultimate source of our ideas is not in the material world or in diagrams, but in our minds, or what Plato calls our "souls."

Plato would have insisted that the drawing in Figure 5-2 does not show squares. He maintained that squares are abstract mathematical ideas. They are quadrilaterals that are parallelograms having a right angle and all sides equal. The ideas represented by the words "quadrilateral," "parallelogram," "side," and "right angle" are all pure mathematical ideas. Today we accept Plato's belief that the elements of geometry are not the imperfect marks that we make on paper. Drawn lines must have width, or thickness, to be seen. Drawn angles and drawn line segments are never perfectly equal in measure.

Plato introduced this view of mathematics when he presented his *concept of an idea.* He wrote in another dialogue, *The Republic*:

> Although mathematicians use visible figures and argue about them, they are not thinking of these figures but of those things which the figures represent; thus it is the square itself and the diagonal itself which are the matter of their arguments, not that which they draw. Similarly, all the figures which they model or draw they use as images.

This statement is analogous to a modern mathematician's view of the relationship between pure and applied mathematics. The use of the word "model" suggests the modern idea of a mathematical model. The latter was not a part of Plato's philosophy, but he probably would have accepted it. Whenever we use mathematics today, we are really making mathematical models of the real world, working with the models, and interpreting the results back in the world of things. If you are told that there are three boys and two girls in the viola section of the school orchestra, you know that there are five viola players, almost without thinking in this case. Your mind automatically abstracted the pure numbers 3 and 2, decided that the mathematical operation should be addition, and thought: "$3 + 2 = 5$," a pure mathematical statement. Then you return to the physical world and thought, "There are 5 violas in the orchestra." In this simple example the model-making process does not really seem to be needed, but it is an important step in more difficult applications.

Plato carried over his theory of ideas to numbers. He was influenced by the Pythagoreans, who thought that 1 should not be called a number. They regarded 1 as a basic unit, from which the whole numbers were made. Although 1 is regarded as a number today, it still has a special place in several mathematical theories. For example, the Peano postulates for constructing the whole numbers use the notion that every whole number has a "successor." A consequence of the postulates is that we obtain the successor of a number by adding 1 to it. Plato made a distinction between "concrete numbers," those used by practical people to count things, and the numbers of pure mathematics. Concrete numbers refer to objects that are not exactly alike. As examples, Plato mentions "camps" (not all camps are alike) and "oxen" (not all oxen are alike). The numbers of pure mathematics, according to him, were made up of pure units, that can be identically the same since they are ideas. The pure unit is considered to be indivisible.

Plato had a great influence on the development of mathematics because he directed the thinking of others toward this subject. He is said to have had a sign over the entrance to his academy: "Let no one ignorant of geometry enter here." In particular, he required the study of mathematics as a preparation for the study of philosophy. Mathematics taught the philosopher to disengage his thinking from the imperfect material world. He could then go on to seek the nature of such absolute qualities of the ideal world as equality, good, and beauty. Again, Plato stressed the importance of the study of mathematics when he was consulted about the Delian problem (see page 97). We even showed a solution to this problem that is attributed to Plato (see page 102). However, it seems unlikely that he devised this solution because it uses a mechanical device, and some people think that Plato, himself, originated the notion that straightedge and compass are the only proper tools for geometric constructions. With these tools lines and circles are drawn that have properties of symmetry and perfection consonant with Plato's philosophy of ideas.

His influence and importance are further shown by the fact that the regular polyhedra are called Platonic solids. These are the convex solids whose edges form congruent regular plane polygons (see Figure 5-3). The most familiar Platonic solid is the cube. Its 6 faces are squares. A remarkable fact about the Platonic solids is that there are exactly five different kinds: the tetrahedron (4 faces), the cube (6 faces), the octahedron (8 faces), the dodecahedron (12 faces), and the icosahedron (20 faces). They were probably known before Plato. Since his time, many interesting problems that involve the construction, the properties, and the uses of Platonic solids have been solved. The thirteenth book of Euclid's *Elements* deals with these solids. At the end of the sixteenth century, Johannes Kepler devised a model for the arrange-

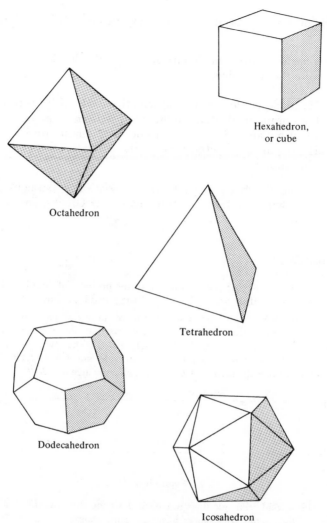

Hexahedron,
or cube

Octahedron

Tetrahedron

Dodecahedron

Icosahedron

Figure 5-3   The five Platonic solids.

ment of the planets which involved concentric spheres inscribed in and circumscribed about Platonic solids. If some of the elements of the definition of the Platonic solids are modified, generalizations are obtained that have new applications. For example, if the polyhedra are not required to be convex, one can form star polyhedra by constructing pyramids on the faces of the Platonic solids. Other generalizations lead to the Archimedean solids and the semiregular solids. Today, properties of polyhedra are important in the study of crystals in chemistry and electronics.

Summary   Plato's chief contributions to the development of mathematics were the following:

1. He made it clear that mathematics deals with ideas, not with lines or numerals that are drawn or written on paper.
2. He stimulated many others to do mathematics for its own sake.
3. He required the study of mathematics as a preparation for philosophy.

His name has been given to the restriction to straightedge and compass in solving the famous problems of the Greeks, and to the regular polyhedra.

EXERCISES 5-2

1 See page 124. Study the slave's solution to the problem of doubling the square. Give a modern proof by use of the method suggested by Figure 5-2.

2 The slave boy's problem was: Given a square, how can you construct a new square having twice the area? Figure 5-4 shows that the problem is easy if we replace "square" by "rectangle." Rectangle $A'B'C'D'$ has twice the area of rectangle $ABCD$ because they have the same altitude and $A'B' = 2AB$. How does this fact prove that the area of $A'B'C'D'$ is twice the area of $ABCD$?

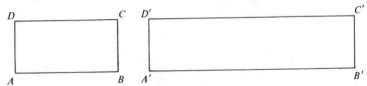

Figure 5-4

The second rectangle does not have the same shape as the first rectangle, however. Two geometric figures have the same shape, or are *similar*, only if, in addition to having their corresponding angles congruent, the ratios of all pairs of corresponding sides are equal. In Figure 5-4, $AD/A'D' = 1$, but $AB/A'B' = \frac{1}{2}$. This suggests the problem: Construct a rectangle that has twice the area of rectangle $ABCD$ and is similar to $ABCD$.

Figure 5-5 suggests a construction for solving this problem. $\overline{AD'}$ is the diagonal of a square whose side is $\overline{AD}$, and $\overline{AB'}$ is the diagonal of a square whose side is $\overline{AB}$. Prove that the area of $AB'C'D'$ is twice the area of $ABCD$ and that these two quadrilaterals are similar.

3 Construct a square having three times the area of a given square. (Hint: See Figure 4-9.)

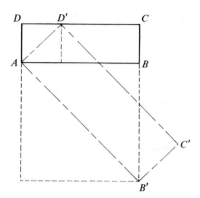

Figure 5-5

4 The lengths of the line segments radiating out from A in Figure 4-32 are $\sqrt{1}$, $\sqrt{2}$, $\sqrt{3}$, $\sqrt{4}$, ... If a curve is drawn through points A, B, C, D, . . . , a spiral is formed that is pleasing to the eye.
(a) Draw a complete turn of this "root spiral."
(b) Draw a rectangle $PQRS$ and use the spiral to construct a rectangle similar to $PQRS$ and with an area that is 5 times the area of $PQRS$.

5 The areas of two circles are proportional to the squares of the lengths of their radii. Given a circle with center M and radius of length r, construct a circle with area equal to
(a) two times the area of the circle (M, r),
(b) three times the area of the circle (M, r),
(c) four times the area of the circle (M, r).

6 Draw each of the following figures, and for each construct a second figure that is similar to the one that you drew and has an area that is
(a) twice as large,
(b) three times as large.
Figures: (1) a square, (2) an equilateral triangle, (3) a scalene traingle, (4) a parallelogram, (5) Figure 5-6.

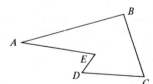

Figure 5-6

7 A popular puzzle is based upon a proof of the Pythagorean theorem. The first part of the puzzle is to assemble four given cardboard congruent quadrilaterals to form a square. The second part of the puzzle is to assemble the same four pieces and a fifth piece, which is a square, to form a third square.

    The secret of the puzzle lies in the way the pieces are constructed. Begin by drawing on stiff paper or cardboard a right triangle, ABC (see Figure 5-7). It is not a part of the puzzle, but it is the underlying secret. Then draw squares

ADEB and *ACHI*. In *ADEB* locate its midpoint, *O*, as the point of intersection of diagonals $\overline{AE}$ and $\overline{BD}$. Through *O* draw $\overline{JK}$, parallel to the hypotenuse $\overline{AC}$, and $\overline{LM}$ perpendicular to it. Cut out the four quadrilaterals *AJOM*, *JDLO*, *LEKO*, and *KBMO*. They are the quadrilaterals of the first part of the puzzle. When they have been cut apart, it is not at all easy to see how to fit them together! The fifth piece, which with the other four are the materials of the second part, is formed by cutting out the square *CBFG*. Figure 5-7 suggests how the five pieces fit together to form a square on side $\overline{AC}$. Cut them out and try to assemble the two squares without looking at the drawing.

Under what circumstances will the quadrilaterals in *ADEB* become triangles? Draw a diagram to represent this case.

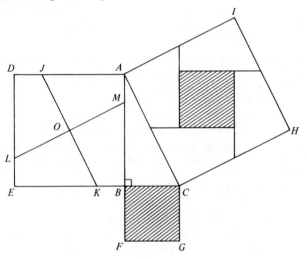

Figure 5-7

8 When you have completed Exercise 7, you have visually verified the Pythagorean theorem for the particular triangle *ABC* that you drew. To *prove* the theorem you must use logical *reasoning*, not depending on the appearance of the quadrilaterals or on how accurately you cut them out. Further, your reasoning must be such that it applies to *every* right triangle, not merely the one that you happened to draw. Give a proof of the Pythagorean theorem based on the dissection of Figure 5-7. (Hint: Prove that the four quadrilaterals in *ADEB* are congruent to each other and that the sum of the lengths of two of their sides equals *AC*, while the difference of the lengths of their other two sides equals *BC*.)

9 The following puzzle shows that cutting apart and reassembling paper figures cannot, by itself, prove a theorem. Draw the square shown in Figure 5-8 and cut it apart into triangles and trapezoids as shown. Reassemble the parts to fit into the rectangle on the right.

Calculate the area of the square and of the rectangle by using the formulas for the area of a square and the area of a rectangle. Explain the resulting paradox. Prove that your explanation is correct.

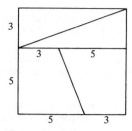

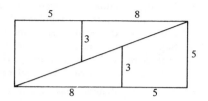

Figure 5-8

10 Show that the measure in degrees of the vertex angles of a regular polygon is

$$\alpha = 180 - \frac{360}{n},$$

where n is the number of sides of the polygon. (Hint: Consider the polygon to be inscribed in a circle, as in Figure 5-9.)

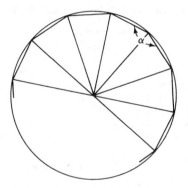

Figure 5-9

11 Equilateral triangles and squares can be arranged to cover a plane in several ways, as shown in Figure 5-10.
   (a) How can regular hexagons be arranged to cover the plane?
   (b) Prove that regular pentagons cannot be arranged to cover the plane.

 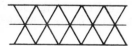 

Figure 5-10

12 If you allow the polygons to be mixed, including regular octagons, hexagons, pentagons, squares, and equilateral triangles, all having sides of the same length, how many different patterns can you find for covering a plane?

13 For a regular polyhedron to be formed, there must be at least three faces meeting at each vertex, and the sum of the measures of the congruent face angles at a vertex must be less than 360°. Why?

(a) Use the formula of Exercise 10 to show that the only regular polygons that can be faces of a regular polyhedron are the equilateral triangle, the square, and the regular pentagon.

(b) Show that there can be no more than five regular polyhedra by showing that there can be three, four, or five equilateral triangles at a vertex, but only three squares or three regular pentagons.

14 Polygons and polyhedra have fascinated men for centuries. *Leonhard Euler* (1707-1783) is credited with the theorem: For any simple polyhedron, the formula $F - E + V = 2$ is true, where F, E, and V are the number of faces, edges, and vertices, respectively. For example, for the cube, $F = 6$, $E = 12$, $V = 8$, and $6 - 12 + 8 = 2$.

(a) The Platonic solids are simple polyhedra (Figure 5-3). Make a table showing the numbers of faces, vertices, and edges for each of them.

(b) Show that these numbers satisfy Euler's formula.

(c) Is Euler's formula valid for a rectangular block through which a square hole has been cut (Figure 5-11a)?

(d) Is Euler's formula valid for a rectangular block that has been beveled at one end after the hole is cut (Figure 5-11b)?

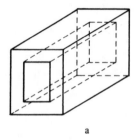
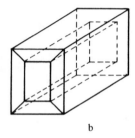

a   b

Figure 5-11

15 Euler's formula can be used to prove that there can be only five regular polyhedra. Look up a proof in reference 27 (Chapter 4).

16 In his dialogue *Timaeus* Plato used the five regular polyhedra to explain scientific phenomena. Look up this explanation and prepare a report.

17 Each of the five regular polyhedra can be constructed from a single sheet of construction paper. An example of one construction for a tetrahedron is suggested by Figure 5-12. Construct a set of regular polyhedra. (Hint: See reference 29 or 30 in this chapter.)

Figure 5-12

18 Exercises 13 and 15 provided different proofs that there *can* be only five regular polyhedra. Prove that there are *exactly* five, or look up a proof in reference 27 (Chapter 4) or 30 (this chapter).

5-3 ARISTOTLE AND HIS THEORY OF STATEMENTS

As we mentioned before, Aristotle, like Plato, was a philosopher and not a mathematician. Plato tried to answer the question, What is the nature of the *objects* of mathematics? Aristotle occupied himself with the problem, What *method* is used in mathematical thinking? Aristotle formulated a theory of proof in which he discussed the part that is played by axioms and definitions. *Eudoxus* (about 360 B.C.) was probably the first to state what we call axioms and to deduce theorems from them. However, Aristotle's writings on the importance of definitions and axioms are the earliest to come down to us.

In the structure of every science, an important part is played by the following three components: statements, concepts, and relations.

An example of a geometrical statement is:

> The sum of the angles of a triangle is equal to
> the sum of two right angles.

In this statement, the two concepts "sum of the angles of a triangle" and "sum of two right angles" are connected by the relation "is equal to."

An example of an arithmetical statement is:

> 4 is greater than 3.

In this statement, the concepts "4" and "3" are connected by the relation "is greater than."

Finally, an example of a statement that is not derived from mathematics is:

> An ape is a mammal.

Here, the concepts "ape" and "mammal" are connected by the relation "is."

In his philosophy, Aristotle developed a theory of statements and a theory of concepts. A theory of relations was not formed until much later.

In mathematics we do not accept the truth of a statement until this statement has been proved, that is, until it has been inferred from other statements whose truth was established earlier. A mathematical statement that can be proved is nowadays called a theorem. Hence, a theorem is proved by deduction from theorems already known. These theorems, however, must in their turn have been proved previously, that is, deduced from preceding theorems. It is clear that we cannot endlessly continue this process of going backward. Somewhere there must be a beginning. From this it follows that *it is necessary to accept some*

*statements as true without supporting the truth of these statements by
a proof.* These statements, which are to be accepted without proof and
serve as starting point for the construction of a mathematical system,
are nowadays called *axioms,* or *postulates.* A science that consists of a
system of theorems which, starting from a number of axioms, are proved
is called a *deductive science.*

Aristotle realized the necessity of starting from some fundamental
truths that must be accepted without proof. He distinguished between
truths that were fundamental to all deductive sciences, which he called
common notions, and *truths that were fundamental to a particular
science,* which he called *special notions.*

Aristotle gave as an example of a common notion: If equals are
subtracted from equals, the remainders are equal. This statement is not
only applicable in the domain of mathematics but in every science that
deals with quantities. The statement that through every two points a
straight line may be drawn is an example of a special notion. The distinction between common and special notions in geometry persisted for
centuries after Aristotle. Common notions were called *axioms* and special notions were called *postulates.* The distinction is no longer observed
today.

Are we allowed to begin a deductive science with some axioms
chosen at pleasure? Not according to Aristotle. He demanded that common notions be self-evident; their truth must be so conspicuous that
nobody doubts it. According to him, the above-mentioned example of
a common notion fulfills this requirement. Aristotle's views of common
and special notions will be further illustrated when we discuss his idea
of concepts in the next section and when Euclid's geometry is discussed
in Chapter 6. A modern view of the role of axioms in a deductive
science will be discussed in Chapter 7.

5-4 CONCEPTS AND DEFINITIONS

In a deductive science we nowadays demand that, except for a few concepts, every concept (and also every relation) be defined; that is, its
meaning must be explained exclusively by means of concepts and relations that have already been introduced. A parallelogram, for instance,
is defined as a quadrilateral whose opposite sides are parallel. We must
therefore know beforehand what is meant by the concepts "quadrilateral" and "side" and by the relations "opposite" and "parallel." Aristotle's analysis of the above definition would have first noted that a
parallelogram is a special type of quadrilateral. He would have said: "A
parallelogram is a quadrilateral with the special property that the opposite sides are parallel." In this case he would have called "quadrilateral"

the *genus proximum* and the special property of "having parallel sides," by which the parallelogram is distinguished from other quadrilaterals, the *differentia specifica.*

With this example we touch the core of Aristotle's theory of definition. Every concept is defined as a subclass of a more general concept. This general concept is called the *genus proximum.* Each special subclass of the *genus proximum* is characterized by special features called the *differentiae specificae.*

In this way, quadrilaterals can be divided into several species, and these into subspecies, and so on. To get a better insight into Aristotle's meaning, we shall divide the set of all convex quadrilaterals into:

1. Parallelograms, quadrilaterals with two pairs of parallel sides.
2. Trapezoids, quadrilaterals of which one pair of sides is parallel and the other pair is not (for a different definition of "trapezoid," see Exercise 4 of Exercises 5-5).
3. Quadrilaterals with no pair of parallel sides.

We divide the parallelograms into:

1. Rectangles, parallelograms with a right angle.
2. Oblique parallelograms, parallelograms without a right angle.

The rectangles can be divided into:

1. Squares, rectangles with all sides congruent.
2. Rectangles whose sides are not all congruent; they are also called oblongs.

The oblique parallelograms can be divided into:

1. Oblique rhombi, oblique parallelograms with all sides congruent.
2. Oblique parallelograms whose sides are not all congruent.

The trapezoids can be divided into:

1. Isosceles trapezoids, trapezoids whose nonparallel sides are congruent.
2. Trapezoids whose nonparallel sides are not congruent.

The diagram on the next page shows the relationships among the concepts just defined. It illustrates a *classification* of the different kinds of quadrilaterals. This classification fulfills the condition that each quadrilateral belongs to exactly one of the subclasses parallelogram, trapezoid, or quadrilateral without parallel sides. Each parallelogram, in turn, is either a rectangle or an oblique parallelogram, and so on. For each *genus proximum*, the subclasses do not overlap and, on the other hand, each element of the *genus proximum* belongs to a subclass. According to Aristotle, a classification must satisfy these requirements.

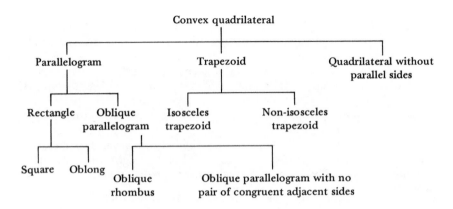

Today, we sometimes find it useful to remove the Aristotelean requirement that the subclasses may not overlap. The Venn diagram in Figure 5-13 illustrates a non-Aristotelean classification of the parallelograms in which the subclass of the rectangles overlaps the subclass of the rhombi to form the subclass of the squares. This arrangement permits us to define the square in the usual way: A square is a quadrilateral that is both a rhombus and a rectangle.

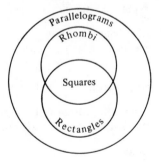

Figure 5-13

Consider again our first classification of quadrilaterals (see page 135). Aristotle would have required that there be at least one quadrilateral in each of the subclasses. That is, it is permissible to separate convex quadrilaterals into parallelograms, trapezoids, and quadrilaterals without parallel sides because there is at least one quadrilateral in each of these subclasses. The following classification of trapezoids would not be permitted, according to Aristotle:

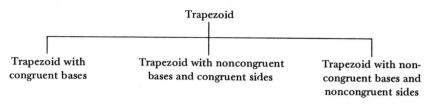

This classification would not have been allowed because, under the first set of definitions, there are no trapezoids with congruent bases (for if two sides of a quadrilateral are both congruent and parallel, it is a parallelogram and not a trapezoid). From a non-Aristotelean viewpoint, such a classification would not be mathematically illegal: it would merely not be very useful, since the set of trapezoids with congruent bases would be the empty set.

To make certain that none of the defined subclasses is empty, *Aristotle required that the existence of a defined object always be proved.* That is why in a Greek mathematical system we never meet a proposition in which a square, for instance, is mentioned unless previously the existence of squares has been proved.

5-5 SPECIAL NOTIONS AND UNDEFINED TERMS

We have seen that a square is a special kind of rectangle, a rectangle with congruent sides. Hence, in order to know what a square is, we must first know what a rectangle is.

A *rectangle* is a parallelogram with a right angle. Hence, to know what a rectangle is, we must first know what a parallelogram is.

A *parallelogram* is a quadrilateral with two pairs of parallel sides. Hence, to know what a parallelogram is, we must first know what a quadrilateral is.

A *quadrilateral* is a figure consisting of four line segments, which are connected in a certain way. Hence, to know what a quadrilateral is, we must first know what a line segment is.

We cannot continue in this way forever. Hence, we must start out with certain *fundamental concepts* that are not explicitly defined. In geometry, for instance, we start from the concepts "point" and "straight line," without defining them.

Nonetheless, Aristotle was of the opinion that it was necessary also to state the *meaning* of these fundamental concepts. According to him, this should be done by way of *statements expressing their essential properties.* In geometry one could choose for this purpose, for instance: "a point has no dimensions," "a straight line has no width and is determined by two of its points."

Aristotle also required that the *existence* of each concept be demonstrated. As an example, we start again with the concept of a square. By definition, a square is a rectangle with congruent sides. Hence, to show the existence of squares we must show that there are rectangles with congruent sides. To be certain that rectangles with congruent sides exist, we must be certain that rectangles exist. A rectangle is a parallelogram with a right angle. Hence, to be certain that parallelograms with a right angle exist, we must be certain that parallelograms exist. And so forth.

We can continue in this way until we come to the fundamental concepts. They, too, must exist, or it would be impossible for the concepts derived from them to exist. However, it is not possible to prove the existence of the fundamental concepts. There is no other choice than to accept their existence without proof.

Aristotle required that for each fundamental concept there be a statement expressing the existence of that concept. For example, in geometry, we must start from such statements as "There exist points" or "There exist straight lines."

Statements that served to lay down the meaning or to assert the existence of the fundamental concepts of a given science are the *special notions* mentioned earlier. Their use was restricted to a given science, such as geometry. Today it is recognized that a deductive system must begin with some undefined terms and some unproved statements using these terms. Some mathematicians choose to regard the postulates of a science as giving "implicit definitions" of the terms used in them. That is, the undefined terms are names for whatever has the properties stated by the postulates.

Summary   According to Aristotle, a deductive science is based on two kinds of statements the truth of which is accepted without proof:

1. Common notions: general truths that are operative in every deductive science.
2. Special notions: truths that underlie a particular deductive science; they are of two kinds: (a) those stating the meaning of the fundamental concepts of that science, and (b) those stating the existence of the fundamental concepts.

All other concepts must be defined. This is accomplished by assigning certain specific properties (*differentiae specificae*) to a concept that is already known (*genus proximum*). The existence of the concepts defined in this way must be proved.

In Chapter 6 it will be shown to what extent Euclid's development of geometry was influenced by Aristotle's philosophy.

EXERCISES 5-5

1 Draw a Venn diagram for the classification on page 135. Be sure to label every region carefully.

2 In the classification on page 135, the parallelograms are separated into the subclasses of the rectangles and the oblique parallelograms. Make a similar classification with the following change: separate the parallelograms into the subclasses of rhombi and nonrhombi.

3 Draw a Venn diagram for the classification in Exercise 2.

4 In the classification on page 135, the convex quadrilaterals are separated into the subclasses: parallelograms, trapezoids, and quadrilaterals without parallel sides. Make a similar classification with the following change: Define a trapezoid to be a convex quadrilateral with *at least* one pair of parallel sides. (There are several possible answers.)

5 Draw a Venn diagram for the classification in Exercise 4.

6 Define a square in the classification of Exercise 2.

7 Define a square in the classification of Exercise 4.

8 Define an oblong in the classification of Exercise 2.

9 Define an oblong in the classification of Exercise 4.

10 The word "isosceles" comes from two Greek words: *iso*, meaning equal, and *skeles*, meaning leg (note: "skeleton" has the same source). There are many other modern words with Greek origins. What mathematical or semimathematical words of this type can you find? Try, for instance, "abacus," "arithmetic," "decagon," "decathlon," "kilometer," "kilogram," "logarithm," "myriad," "pentagon," "pentathlon." Use a dictionary that gives the origins of words, such as *The Oxford English Dictionary*.

11 Aristotle is supposed to have been the author of a paradox that is still occasionally proposed as a mathematical puzzle. We will state Aristotle's paradox in modern terms.

Figure 5-14 represents the wheel of an automobile. The tire rests upon

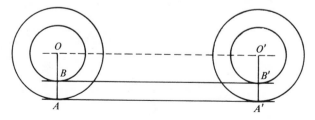

Figure 5-14

the road at A and the hubcap just touches the curb at B. Now let the car roll forward for just one revolution of the wheel. The radius $\overline{OBA}$ will complete one revolution about the center O and come to rest in the new position $O'B'A'$. Clearly, the length AA' is then equal to the circumference of the tiré. But the hubcap has also completed one revolution as it moved along the curb. Hence, BB' is the circumference of the hubcap, or of the circle with radius $\overline{OB}$. In other words, the circle with radius $\overline{OB}$ has the same circumference as the circle with radius $\overline{OA}$, since $\overline{AA'}$ and $\overline{BB'}$ are congruent line segments! Of course, we know that circles whose radii are of different length have different circumferences. What is the explanation of this paradox?

REFERENCES

For more about Greek philosophers, see references 7 (Chapter 1) and 19 and 20 (Chapter 3).

There is more about Euler's theorem on polyhedra in reference 27 (Chapter 4). This reference includes a proof of the theorem and an application of the theorem to a proof that there are only five Platonic solids.

For models of polyhedra, see

[29] Wenninger, Magnus J., *Polyhedron Models for the Classroom*. Washington, D.C.: National Council of Teachers of Mathematics, 1966. The Platonic solids are pictured on page 2.
[30] Wenninger, Magnus J., *Polyhedron Models*. New York: Cambridge University Press, 1970.

6

EUCLID

6-1 ELEMENTS

In our time we know mathematics as a system of axioms, definitions, and propositions. We inevitably move toward such a system if we try (as Plato required) to free mathematics from material things and hence from concrete figures. For if an empirical demonstration of the truth of propositions is no longer permitted (for example, by reading them from a figure), there is no other choice but to prove them by logical reasoning. This means that the propositions must be deduced from propositions already proved and, thus going back, ultimately from the axioms.

Another consequence of the logical separation of mathematics from the material world is that the meaning of a new concept may not be demonstrated with the aid of a figure only, but that for every new concept an exact definition must be given. (Of course, in developing or in teaching mathematics, we depend on the motivations that come from the real world and on the clarity that comes from diagrams.)

The Greeks were convinced that mathematics should be free from empirically acquired knowledge. They accepted the consequences of this belief and built up geometry in the way described above. To a system of mathematical propositions based on axioms they gave the name "Elements." According to Proclus, a set of Elements was compiled by Hippocrates about 100 years before Euclid. After him, several other

Greek mathematicians drew up similar systems. The oldest Elements whose contents have been preserved are the *Elements* of Euclid.

Little is known of *Euclid* (about 300 B.C.) himself. He lived in the time of the first Ptolemy in Alexandria, where he taught mathematics at the university, called the Museum. No important discoveries are attributed to him, but from his work we may infer that he was an excellent teacher. He set forth the principles of mathematics in such a way that they became understandable to students of every time. The contents of his work were, for the greater part, taken from his predecessors. We find in Euclid's *Elements* the results of the work of, among others, Theaetetus and Eudoxus. It is probably because of its great usefulness that the *Elements* have been preserved. However, as happened with all Greek works, the original work has been lost. It was copied countless times and provided with commentaries; later writers introduced changes wherever they thought them desirable. It is therefore no longer possible to reconstruct the original text in every detail. In the Middle Ages, the *Elements* became known in Western Europe via the Arabs and the Moors. There the *Elements* became the foundation of mathematical education. More than 1000 editions of the *Elements* are known. In all probability it is, next to the *Bible*, the most widely spread book in the civilization of the Western world.

6-2 THE STRUCTURE OF THE *ELEMENTS* OF EUCLID

According to Plato, mathematical knowledge can be acquired only by reasoning. Therefore, no properties should be read from the figure, but an exact proof must be given of every property, that is, a proof in which the figure is not used. Euclid has tried to keep close to this requirement.

According to Aristotle, in constructing a mathematical system we must start from *common notions*, which underlie all deductive thinking. Over and above we must start from *special notions*, in which the existence of the fundamental concepts of mathematics is postulated or in which their meaning is stated. Finally, the other concepts must be defined with reference to a *genus proximum* and the *differentiae specificae*, and the existence of the defined concepts must be proved. We shall see that Euclid has tried to construct his system in accordance with the Aristotelean prescriptions.

The *Elements* consists of 13 books. In the first six books, plane geometry is discussed. The extent of the treated subject matter corresponds more or less with what is taught in a modern high school course. In the next three books, number theory is developed. In Book X, irrational ratios are discussed. Finally, Books XI through XIII deal with solid geometry. We shall discuss most of Book I and some parts of the remaining books.

6-3 THE DEFINITIONS

The first book begins with a list of 23 definitions. Some of them follow below.

I   *A* point *is that which has no part.*
II   *A* line *is breadthless length.*
III *The extremities of a line are* points.
IV   *A* straight line *is a line that lies evenly with the points of itself.*
V   *A* surface *is that which has length and breadth only.*
VI   *The extremities of a surface are* lines.
VII *A* plane surface *is a surface that lies evenly with the straight lines on itself.*

Before taking a closer look at these definitions, we remind the reader of the way in which, according to Aristotle, concepts must be introduced. Aristotle distinguishes two kinds of concepts: fundamental concepts and concepts derived from them.

a. The fundamental concepts cannot be defined. Their essential properties are formulated in special notions.
b. The remaining concepts are defined, starting from the fundamental concepts. First, there is a concept that is already known, the *genus proximum*. Those particular cases of the *genus proximum* which satisfy certain requirements, the *differentiae specificae*, constitute the new concept (for instance, *genus proximum*: triangle; *differentia specifica*: two sides are equal; new concept: isosceles triangle).

Hence, every deductive science has to start out with a number of fundamental concepts, and their meaning must be laid down in special notions. Therefore, the seven "definitions" given above are not proper definitions, but they can be considered to be special notions which state the meaning of the fundamental concepts appearing in them. These fundamental concepts are: point, line, straight line, surface, and plane.
    The first definition describes what has to be understood by a point: "A point is that which has no part." Hence, a point should not be conceived of as a small dot but as something that has no dimension at all and which therefore is immaterial. From the second definition it appears that a line, too, is immaterial. It is not a thin thread but "breadthless length." Similarly, for a surface.
    The meaning of Definitions IV and VII is vague. What does it mean that a straight line "lies evenly with the points of itself"? Euclid may have thought of sighting along a rod and concluding that it is straight if no point sticks out, that is, if the rod is seen as a single point. Similarly,

Definition VII would then mean that a plane surface is seen as a straight line when one looks along it.

Euclid now continues with a series of definitions in which the meaning of the derived concepts is explained.

VIII   *A* plane angle *is the inclination to one another of two lines in a plane that meet one another and do not lie in a straight line.*

IX   *And when the lines containing the angle are straight, the angle is called* rectilineal.

X   *When a straight line set up on a straight line makes the adjacent angles congruent to one another, each of the congruent angles is* right, *and the straight line standing on the other is called a* perpendicular *to that on which it stands.*

In Definition VIII the angle between two lines is defined. These lines need not be straight lines, for the rectilineal angle is not defined until Definition IX. Definition VIII is not yet formulated in the way prescribed by Aristotle. The *"inclination* to one another of two lines" is not a *genus proximum* that has been defined previously.

Definition IX is the first that *meets the Aristotelean requirements* of a definition. According to this definition, a *rectilineal angle* is an angle that is formed by two straight lines. Here, "angle" is the *genus proximum.* The special property that makes the rectilineal angles a special case of angles is that the two lines forming the angle are straight lines. This special property is therefore the *differentia specifica.*

Also, Definition X is in agreement with Aristotle's requirements. What is a *right angle*? A special kind of rectilineal angle (*genus proximum*). What special property does the right angle have? The property that, after producing one of the two lines, two congruent angles are formed (*differentia specifica*).

Definition X is followed by definitions of an obtuse and an acute angle, a circle, the center and a diameter of a circle, a semicircle, a polygon, and several kinds of triangles and quadrilaterals. The definition of a circle is as follows:

XV   *A* circle *is a plane figure contained by one line such that all the straight lines falling upon it from one point among those lying within the figure are congruent to one another.*

The list of definitions ends with:

XXIII   Parallel straight lines *are straight lines which, being in the same plane and being produced indefinitely in both directions, do not meet one another in either direction.*

With the term "straight line" Euclid evidently does not think of an infinite straight line but of a line segment (see Definition III: the extremities of a line are points). In the present definition he thus calls two coplanar line segments parallel when they have no point in common, even when extended.

EXERCISES 6-3

1 Indicate those definitions listed in this section which meet Aristotle's requirements for a definition.

2 Euclid could not avoid using words that had not been previously defined in his definitions. Sometimes they were key words, such as "inclination" in Definition VIII. Find other undefined key words in the definitions listed in this section.

3 Try to make up a "better" definition than Euclid gives of a point, a line, a straight line, and a plane.

4 In which definitions listed in this section does it appear that Euclid speaks of straight lines when he means line segments?

5 In which definitions listed in this section does Euclid speak of lines when he does not necessarily mean straight lines?

6-4 POSTULATES AND COMMON NOTIONS

Euclid bases his *Elements* on five *postulates*. We shall first treat Postulates I through III (the word "and," with which some of the postulates begin, means "let it be postulated").

POSTULATES

I *Let the following be postulated: to draw a straight line from any point to any point.*

II *And to produce a finite straight line continuously in a straight line.*

III *And to describe a circle with any center and distance.*

Postulate I says that any point can be connected with any other point by means of one line segment.

Postulate II means that a line segment can always be extended. As a result of the extension, a longer line segment is generated. According to the same postulate, this longer line segment can again be produced, and so on. The content of Postulate II therefore amounts to this: there are infinite straight lines.

We see that the term "straight line" in the postulates corresponds with our concept "line segment." The concept "straight line" in the sense of an infinite straight line does not occur in the postulates, but its existence is implicitly postulated in II. In modern terminology the contents of Postulates I and II would therefore be summed up by: *Through any two points, one straight line can be drawn; a straight line is infinite in extent.*

It is stated in Postulate III that there are circles and that a circle is determined when we know its center and its radius.

Postulates I through III can be compared with Aristotle's special notions. According to Aristotle, the existence of the fundamental concepts must be postulated in special notions. Postulates I through III have to be seen as special notions in which the existence of the (modern) fundamental concepts line segment, straight line, and circle is stated. It is true that the existence of points is not stated explicitly, but evidently, Euclid accepted their existence.

Postulates IV and V read:

IV   *And that right angles are congruent to one another.*

V   *And that, if a straight line falling on two straight lines makes the interior angles on the same side less than two right angles, the two straight lines, if produced indefinitely, meet on that side on which are the angles less than the two right angles.*

These two postulates differ so strikingly from the first three that the question arises why Euclid put all five in one list. Postulates I through III are existence postulates, hence special notions in the Aristotelean sense. *This does not apply to Postulates IV and V.* In Postulate IV, the existence of right angles is not stated but we are required to assume that all right angles are congruent. In Postulate V, we are required to accept that two straight lines which meet certain conditions have a point of intersection. Is it in agreement with Aristotle's views to make such demands? Not at all. For Postulates IV and V are not common notions, because they refer exclusively to geometry. And they are not special notions either, for their purpose is not to establish the meaning or the existence of certain fundamental concepts of the geometrical system.

If Euclid had kept strictly to the Aristotelean prescriptions, he would have left out Postulates IV and V. Why then did he add them? The answer is simple. He did not see a way to build up his system of plane geometry without accepting Postulates IV and V. Neither did he see a way to demonstrate their truth by means of a proof. So he had no other choice but to accept their truth. And because they are specifically geometrical properties (not applicable outside geometry) which are accepted without proof, it was reasonable to list them with the postulates.

Thus we find two kinds of postulates in Euclid's list:

1. *Existence postulates, in which the existence of certain fundamental*

concepts is assumed (I through III).
2. *Postulates in which it is assumed that geometrical figures have certain specific properties (IV and V).*

We recall Aristotle's requirement that a deductive science be based upon (in addition to postulates) *common notions*, or *axioms*. As we know, these common notions do not underlie just one particular science but form the basis of all deductive thinking. In agreement with Aristotle, Euclid also starts from common notions. They are the following.

COMMON NOTIONS

    I   *Things that are equal to the same thing are equal to one another.*

    II   *If equals be added to equals, the wholes are equal.*

    III   *If equals be subtracted from equals, the remainders are equal.*

    VII   *Things that coincide with one another are equal to one another.*

    VIII   *The whole is greater than the part.*

The irregular numbering stems from the fact that some of the common notions that appeared in the commentaries were later deleted because they did not belong to the original text of the *Elements*.

Starting from the foundations, Euclid now establishes the geometrical structure of the *Elements*. In doing so, he tries to satisfy Aristotle's requirements:

1. Each new statement must be proved.
2. Each new notion must be defined; moreover, its existence must be proved.

Book I of the *Elements* contains 48 propositions. They deal with congruence of triangles, parallel lines, and area. The book terminates with the Pythagorean theorem and its converse. We shall discuss the main contents of Book I, considering the Pythagorean theorem as our ultimate goal, and selected parts of the other books.

6-5   THE MEANING OF A CONSTRUCTION

PROPOSITION 1   *To construct on a given line segment $\overline{AB}$ an equilateral triangle.*

   Construction   The solution of this simple problem amounts to the following (see Figure 6-1).

   1. Draw the circles $(A, \overline{AB})$ and $(B, \overline{BA})$. (We use the notation

$(A, \overline{AB})$ to denote a circle with the point A as its center and the line segment $\overline{AB}$ as a radius.)

2. Let C be a point of intersection of the circles.

3. Draw the line segments $\overline{AC}$ and $\overline{BC}$.

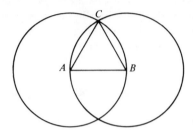

Figure 6-1

Then ABC is an equilateral triangle that is constructed on the line segment $\overline{AB}$.

Proof

$\overline{AC} \cong \overline{AB}$ (radii of a circle are congruent: Definition XV).

$\overline{BC} \cong \overline{AB}$ (Definition XV).

From this it follows that

$$\overline{AC} \cong \overline{BC} \text{ (Axiom I).}$$

Hence, $\triangle ABC$ is an equilateral triangle.

Remarks

1. The construction and the proof of Proposition 1 as stated above are not a literal translation of Euclid's text. However, we followed as closely as possible his line of thought, but we used a modern notation and mode of expression. In particular, we used the word "congruent" and the symbol "$\cong$" where Euclid talks of "equality." This use of the word "congruent" and the symbol "$\cong$" should not be confused with their modern use, for Euclid does not assign a length to a line segment or a measure to an angle. Euclid's notion of equality of line segments and angles must be interpreted as meaning that equal line segments and equal angles can be moved in the plane until they coincide.

We shall continue to use the term "line segment" where Euclid speaks of a straight line, and we shall use the words "line" and "straight line" to indicate an infinite straight line. In these and other respects we shall often use a modern way of expressing the contents of Book I of the *Elements*.

2. See steps 1 and 3 of the construction. It is possible to draw the circles $(A, \overline{AB})$ and $(B, \overline{BA})$ on the strength of Postulate III. These circles therefore exist. The drawing of $\overline{AB}$ and $\overline{BC}$ can be done on the

strength of Postulate I. This again means: $\overline{AC}$ and $\overline{BC}$ exist.

3. In step 2 of the construction, one of the points of intersection of the circles $(A, \overline{AB})$ and $(B, \overline{BA})$ is constructed. We say: "let C be a point of intersection," but we mean: "there exists a point of intersection C."

But how does Euclid know that the circles have a point of intersection? Is there a postulate in which the existence of such a point is guaranteed? It is clear that this is not so. Could we perhaps prove the existence of a point of intersection on the basis of what we know so far, hence on the strength of the postulates and the common notions alone? Have a try. You will soon see that it cannot be done. The existence of a point of intersection seems sensible when we look at the figure, but *to prove the existence, we must have more postulates at our disposal than those stated by Euclid.* Euclid probably overlooked this.

4. When Euclid finally remarks that a triangle has been constructed that fulfills the requirements, he means that he has proved the existence of this triangle.

5. It should now be clear that every construction is a proof of a special kind, a proof in which the existence of a certain figure is shown. Perhaps this is the reason Euclid lists the constructions in sequence with the propositions and treats them as equivalent.

The preceding remarks show that Proposition I and its proof can be worded in the following two ways.

PROPOSITION 1 *To construct on a given line segment $\overline{AB}$ an equilateral triangle.*

Construction

1. *Draw the circles $(A, \overline{AB})$ and $(B, \overline{BA})$.*
2. *Let C be a point of intersection.*
3. *Draw the line segments $\overline{AC}$ and $\overline{BC}$.*

Triangle ABC is equilateral and has $\overline{AB}$ as a side. Thus, a triangle has been *constructed* that fulfills the requirements.

PROPOSITION 1 *There is an equilateral triangle that has a given line segment $\overline{AB}$ as side.*

Proof

1. The circles $(A, \overline{AB})$ and $(B, \overline{BA})$ *exist.*
2. There *exists* a point of intersection; we indicate it by C.
3. The line segments $\overline{AC}$ and $\overline{BC}$ *exist.*

Triangle ABC is equilateral and has $\overline{AB}$ as a side. Thus, the *existence* of a triangle that fulfills the requirements has been proved.

Thus, a construction proves to be in essence an existence proof. Euclid shows which figures have to be thought of to obtain the required equilateral triangle and proves at the same time that this is possible on the strength of the postulates.

Why is it that Euclid says: "*draw* circle $(A, \overline{AB})$" and "*draw* line segment $\overline{AC}$"? After all, he would have been more correct in saying "circle $(A, \overline{AB})$ exists" and "line segment $\overline{AC}$ exists," or "think of circle $(A, \overline{AB})$" and "think of line segment $\overline{AC}$." In this simple case that would indeed have been more reasonable. But in more complicated cases, the reasoning would be hard to follow if it consisted of a lengthy enumeration of circles and straight lines that we have to think of successively. To support our imagination we draw a figure. *However, this figure is only an expedient to enable us to follow the reasoning more easily,* a necessary evil, as it were. For the sake of convenience, we adapt our terminology to it. For this reason Euclid uses (just as we do today) the phrases "draw circle $(A, \overline{AB})$" and "draw line segment $\overline{AC}$."

We therefore distinguish two kinds of terminology:

1. One in which we speak of the *existence* of lines and circles.
2. One in which we speak of the *drawing* of lines and circles.

In terminology 1 we make use of the *existence postulates*; these are the "instruments" by means of which the existence of a figure is proved.

In terminology 2 we make use of the *straightedge* and *compass*; these are the instruments by means of which the corresponding visible figure is drawn.

Whenever we say that we are drawing a straight line or a circle (with *straightedge* and *compass*), what we really mean is that we are applying an *existence postulate*. That is why Euclid proves the existence of a figure by showing the possibility of constructing the figure. *This construction must fulfill the requirement that no other instruments be used but straightedge and compass, for in the existence postulates only the existence of straight lines and circles is postulated.*

6-6   THE PURPORT OF POSTULATE III

PROPOSITION 2   *Given a point P and a line segment $\overline{AB}$, construct a point X such that $\overline{PX} \cong \overline{AB}$.*

Construction   (See Figure 6-2.)

1. Draw $\overline{AP}$ (Postulate I).
2. Construct the equilateral triangle APC (Proposition 1).
3. Draw circle $(A, \overline{AB})$ (Postulate III).
4. Extend $\overline{CA}$ (Postulate II) and let D be the point of intersection of circle $(A, \overline{AB})$ and $\overline{CA}$ extended.
5. Draw a circle $(C, \overline{CD})$ (Postulate III).
6. Extend $\overline{CP}$ (Postulate II) and let X be the point of intersection of circle $(C, \overline{CD})$ and $\overline{CP}$ extended.

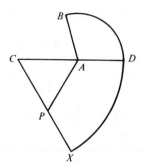

Figure 6-2

Then X is the required point.

Proof   (" $\therefore$ " is an abbreviated way to write "hence" or "therefore.")

$\overline{CX} \cong \overline{CD}$ (Definition XV) $\Big\}$ $\therefore$ $\overline{PX} \cong \overline{AD}$ (Axiom III).
$\overline{CP} \cong \overline{CA}$ (by construction) $\Big)$

We further have

$$\overline{AB} \cong \overline{AD} \quad \text{(Definition XV),}$$

and hence

$$\overline{PX} \cong \overline{AB} \quad \text{(Axiom I).}$$

Remarks

1. See steps 4 and 6. Just as with Proposition 1 we notice here a gap in the method of proof. By what right does Euclid assume that (in 4) circle $(A, \overline{AB})$ intersects $\overline{CA}$ extended and (in 6) that circle $(C, \overline{CD})$ intersects $\overline{CP}$ extended? Euclid here makes use of the fact that a ray which has the center of a circle as its endpoint has a point in common with the circle. However, he gives no proof of the existence of such a point. He could not have done so anyway without adding appropriate postulates. He therefore concludes on intuitive grounds that the existence of a point of intersection has been established.

2. Postulate III says that with any point as center and any radius, a circle can be drawn. Could Euclid then not have proved Proposition 2 simply by saying: Draw a circle with center P and radius $\overline{AB}$? He does not do this. Apparently, he does not want Postulate III to be understood in such a way, that it is possible to draw a circle with center P and a line segment congruent to $\overline{AB}$ ($P \neq A$, $P \neq B$) as a radius. But he does draw several circles with a given center and a radius that is a line segment with the given center as one of its endpoints. Thus, it appears to be his intention that circles can only be drawn in this way. Hence, Postulate III should be read as follows: To draw a circle with any point P as its center and any line segment $\overline{PA}$ as a radius. That is, the given

radius must be a line segment with the given center as one of its endpoints.

3. The preceding remark can be illustrated by imagining an instrument that will allow us to draw only those circles which can be constructed as a consequence of Postulate III. Such an instrument is a compass having the property of collapsing as soon as either leg is lifted from the paper. With such a compass we can draw circle $(P, \overline{PA})$ but not circle $(P, \overline{AB})$, for when we would try to move $\overline{AB}$ to P, the compass would collapse.

4. Proposition 2 shows that it is possible to "move" any line segment $\overline{AB}$ in such a way that one of its endpoints coincides with P. From this it follows that it is also possible to draw immediately a circle with center P and radius $\overline{AB}$. Hence, after Proposition 2 has been proved, the "collapsible compass" and the (rigid) compass are completely equivalent in their use.

In Exercises 2 through 8 of Exercises 6-6 the reader will be asked to carry out a series of constructions with the collapsible compass, which will result in a construction that can be used to give another proof of Proposition 2.

PROPOSITION 3   *Given two unequal line segments, to lay off on the greater a line segment equal to the smaller.*

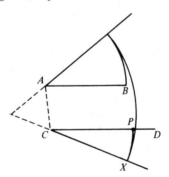

Figure 6-3

Construction   (See Figure 6-3.) Let $\overline{AB}$ and $\overline{CD}$ be the given line segments. Assume that $\overline{CD} > \overline{AB}$.

1. Construct a line segment $\overline{CX}$ that is congruent to $\overline{AB}$ (Proposition 2).
2. Draw circle $(C, \overline{CX})$ (Postulate III).
3. Let P be the point of intersection of this circle and $\overline{CD}$. Then $\overline{CP}$ is the required line segment.

Proof

$$\overline{CX} \cong \overline{AB} \quad \text{(by construction)}$$
$$\overline{CP} \cong \overline{CX} \quad \text{(Definition XV)}$$

$\left.\begin{matrix} \\ \\ \end{matrix}\right\}$ $\therefore$ $\overline{CP} \cong \overline{AB}$ (Axiom I).

EXERCISES 6-6

1 Verify that the construction in Proposition 1 (page 147) can be carried out with straightedge and collapsible compass.

Carry out the constructions in Exercises 2 through 8 by using your compass as if it were a collapsible compass. Give a proof of each construction, stating the steps and the reasons. Do not use Propositions 2 or 3 as a reason.

2 On one of the sides of a given $\angle A$ the two points P and Q are given. Construct on the other side two points R and S such that $\overline{RS} \cong \overline{PQ}$.

3 Given: a line segment $\overline{AB}$ and a straight line l. Construct on l two points C and D such that $\overline{CD} \cong \overline{AB}$. Consider the following cases:
(a) l intersects $\overline{AB}$.
(b) l intersects the extension of $\overline{AB}$.
(c) $l \parallel \overline{AB}$.

Remark From the solution of Exercise 3, we note that on a given line a segment can be constructed congruent to a given segment (by using straightedge and collapsible compass).

4 Given: an equilateral triangle ABC and a point P on $\overline{AB}$ extended (B between A and P). Construct on $\overline{CA}$ extended a point Q (A between C and Q) such that $\overline{AQ} \cong \overline{BP}$.

5 Given: an equilateral triangle ABC and a point P on side $\overline{AB}$. Construct on side $\overline{AB}$ a point Q such that $\overline{AQ} \cong \overline{BP}$.

6 Given: a line segment $\overline{AB}$ and a point P on $\overline{AB}$. Construct on $\overline{AB}$ a point Q such that $\overline{AQ} \cong \overline{BP}$.

7 On a straight line l the points A, B, and C are given such that B lies between A and C. Construct a point X on l such that $\overline{AX} \cong \overline{BC}$.

Remark From the solution of Exercise 7, we note that, given a straight line l, a point P on l, and a segment $\overline{AB}$ on l, we can construct on l a segment with endpoint P, congruent to $\overline{AB}$ (by using straightedge and collapsible compass). In connection with the remark following Exercise 3 we are now able to carry out the following: Given a straight line l, a point P on l, and a line segment $\overline{AB}$ that does not lie on l, to construct on l a segment with endpoint P, congruent to $\overline{AB}$ (by using straightedge and collapsible compass).

8 Given: a straight line *l*, a point *P* on *l*, and a line segment $\overline{AB}$ that does not lie on *l*. Construct a point *X* on *l* such that $\overline{PX} \cong \overline{AB}$.

Remark The solution of Exercise 8 gives a construction for Proposition 3 that is different from the one given by Euclid.

9 Find a construction for Proposition 2, by using only a collapsible compass, no straightedge.

6-7 CONGRUENCE

PROPOSITION 4 *Two triangles are congruent if two sides of one triangle are congruent with two sides of the other and the included angles are congruent.*

Proof This is demonstrated by showing that one of the triangles can be placed on the other such that it fits exactly.

Remarks

1. In the proof of Proposition 4, Euclid *moves* a triangle to another place. What basis does he have for such a motion? Neither the common notions, nor the postulates, nor the preceding propositions provide a basis for it since the concept of "motion" is not mentioned in them. Here again we meet a deficiency in the structure of Euclid's system. *The concept of "motion" is inferred from our experience with material things.* However, as soon as we want to apply this concept to mathematics, we must provide an axiomatic basis for it. Euclid failed to do so. Was Euclid aware of this gap in the construction of his system? No doubt. In the proof of Proposition 2 he went to a lot of trouble to show how line segment $\overline{AB}$ can be transferred to a given point *P*. In that case he properly realized that it was not admissible to say without comment: Move the line segment $\overline{AB}$ such that *A* falls upon *P*. He also must have understood that there was no justification for proving Proposition 4 by means of the transfer of a triangle, but he saw no other way out.

2. The wording of Proposition 4 (just as the wording of some other propositions) is a free translation of Euclid's text. For comparison, a strict translation of the original text reads: "If two triangles have two sides equal to two sides, respectively, and have the angles contained by the equal straight lines equal, they will also have the base equal to the base, the triangle will be equal to the triangle, and the remaining angles will be equal to the remaining angles, respectively, namely those which the equal sides subtend."

The congruence (equality) of the triangles, the bases, and the other angles mentioned in this statement follows from Axiom VII. For, since the triangles can be made to coincide, they are congruent, and, for the same reason, their corresponding parts are congruent.

PROPOSITION 5 *In an isosceles triangle the base angles are congruent; also the angles formed by the base and the extensions of the sides are congruent.*

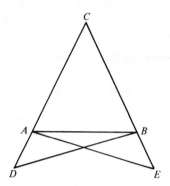

Figure 6-4

Proof Let ABC be a triangle in which $\overline{CA} \cong \overline{CB}$ (see Figure 6-4). Choose on the extension of $\overline{CA}$ a point D (Postulate II) and construct E on the extension of $\overline{CB}$ such that $\overline{CE} \cong \overline{CD}$ (Postulate III). Draw $\overline{AE}$ and $\overline{BD}$ (Postulate I). Then

$$\left.\begin{array}{l} \overline{CD} \cong \overline{CE} \\ \overline{CB} \cong \overline{CA} \\ \angle C \cong \angle C \end{array}\right\} \quad \therefore \quad \triangle CBD \cong \triangle CAE \quad \text{(Proposition 4).}$$

Hence,

$$\overline{BD} \cong \overline{AE} \tag{1}$$
$$\angle CBD \cong \angle CAE \tag{2}$$
$$\angle CDB \cong \angle CEA. \tag{3}$$

From $\overline{CD} \cong \overline{CE}$ and $\overline{CA} \cong \overline{CB}$, it follows that

$$\overline{AD} \cong \overline{BE} \quad \text{(Axiom III).} \tag{4}$$

From statements (1), (3), and (4) it follows that

$$\triangle AEB \cong \triangle BDA \quad \text{(Proposition 4).}$$

Hence,

$$\angle ABE \cong \angle BAD. \tag{5}$$

This proves the second half of the proposition. From the congruence of $\triangle AEB$ and $\triangle BDA$ it further follows that

$$\angle BAE \cong \angle ABD. \tag{6}$$

From statements (2) and (6) it follows that

$$\angle CAB \cong \angle CBA \quad \text{(Axiom III).}$$

This proves the first half of the proposition. Therefore, Proposition 5 has been proved.

Remark   In this proof one might be tempted to say immediately after statement (5): "Hence, also the supplements of the angles *ABE* and *BAD* are congruent." Then, however, we would make use of the theorem "congruent angles have congruent supplements." This is not admissible, for this theorem has not yet been proved; it is Proposition 13.

The converse of Proposition 5 is:

PROPOSITION 6   *If two angles of a triangle are congruent, then the sides opposite these angles are congruent.*

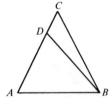

C
D
A   B   **Figure 6-5**

Proof   Let *ABC* be a triangle in which $\angle CAB \cong \angle CBA$ (see Figure 6-5). We must prove that $\overline{AC} \cong \overline{BC}$.

Assume that $\overline{AC} \ncong \overline{BC}$; one of $\overline{AC}$ and $\overline{BC}$ is then the greater. Let this be $\overline{AC}$. On $\overline{AC}$, let $\overline{AD} \cong \overline{BC}$ (Proposition 3).

Now we have

$$\left.\begin{array}{l} \overline{AB} \cong \overline{BA} \\ \overline{AD} \cong \overline{BC} \\ \angle DAB \cong \angle CBA \end{array}\right\} \quad \therefore \ \triangle DAB \cong \triangle CBA \quad \text{(Proposition 4).}$$

Hence, $\angle DBA \cong \angle CAB$. But we also have $\angle CAB \cong \angle CBA$. From this it follows that

$$\angle DBA \cong \angle CBA \quad \text{(Axiom I).}$$

Then, however, the "whole," $\angle CBA$, would be equal to the "part," $\angle DBA$, in contradiction to Axiom VIII. So the assumption $\overline{AC} \ncong \overline{BC}$ is not correct. Hence, $\overline{AC} \cong \overline{BC}$.

Remark   In the Elements, Axiom VIII often plays a role in an indirect proof. The proof of Proposition 6 is an example of this.

PROPOSITION 7 _If in the triangles ABC and ABD, which are on the same side of_ $\overline{AB}$, $\overline{AC} \cong \overline{AD}$ _and_ $\overline{BC} \cong \overline{BD}$, _then C coincides with D._

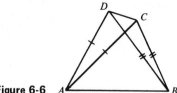

Figure 6-6

Proof (See Figure 6-6.) Assume that *C* does not coincide with *D*. Draw $\overline{CD}$ (Postulate I). Then

$$\angle ACD \cong \angle ADC \quad \text{(Proposition 5)}$$
$$\angle ADC > \angle BDC \quad \text{(Axiom VIII)} \quad \Big\} \quad \therefore \angle ACD > \angle BDC.$$

We then further have

$$\angle ACD > \angle BDC$$
$$\angle BCD > \angle ACD \quad \text{(Axiom VIII)} \quad \Big\} \quad \therefore \angle BCD > \angle BDC.$$

But according to Proposition 5, we also have $\angle BCD \cong \angle BDC$, which is impossible. Hence, *C* does coincide with *D*.

Remarks

1. This proof is not quite complete. Also, other positions of the point *D* with regard to triangle *ABC* are possible. For instance, *D* may lie inside the triangle. *If there are several possibilities, Euclid gives the proof of one case only* and, as a rule, of the most difficult one. (See Exercise 3 of Exercises 6-7.)

2. We note that in the proof two properties of inequalities are used. The first of these properties is:

If $\angle A \cong \angle B$ and $\angle B > \angle C$, then $\angle A > \angle C$.

The second is:

If $\angle A > \angle B$ and $\angle B > \angle C$, then $\angle A > \angle C$.

Euclid has not included axioms by which these conclusions are justified. Today, we would call such axioms transitivity axioms.

PROPOSITION 8 _If the three sides of one triangle are congruent to the sides of another triangle, then the triangles are congruent._

Proof The proof is given by placing one of the triangles on the other such that two of their corresponding vertices (and therefore, their bases) coincide and that their third vertices lie on the same side of the common base. According to Proposition 7, these third vertices then coincide, from which it follows that the triangles are congruent.

PROPOSITION 9   *To bisect a given angle.*

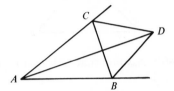

Figure 6-7

Construction   Let ∠A be the given angle (see Figure 6-7).

1. Determine on the sides the points B and C such that $\overline{AB} \cong \overline{AC}$ (Postulate III).
2. Construct an equilateral triangle BCD (Proposition I).
3. Draw $\overline{AD}$ (Postulate I).

Then $\overline{AD}$ bisects the given angle.

Proof

$$\left.\begin{array}{c} \overline{AC} \cong \overline{AB} \\ \overline{CD} \cong \overline{BD} \\ \overline{AD} \cong \overline{AD} \end{array}\right\} \therefore \triangle ADC \cong \triangle ADB \quad \text{(Proposition 8).}$$

From this it follows that ∠CAD ≅ ∠BAD.

PROPOSITION 10   *To bisect a given line segment.*

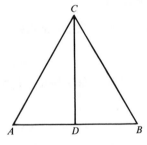

Figure 6-8

Construction   Let $\overline{AB}$ be the given line segment (see Figure 6-8).

1. Construct an equilateral triangle ABC (Proposition I).
2. Bisect ∠ACB (Proposition 9).

Then the point of intersection, D, of the bisector and $\overline{AB}$ is the midpoint of $\overline{AB}$.

Proof

$$\left.\begin{array}{c} \overline{AC} \cong \overline{BC} \\ \angle ACD \cong \angle BCD \\ \overline{CD} \cong \overline{CD} \end{array}\right\} \quad \therefore \ \Delta ACD \cong \Delta BCD \quad \text{(Proposition 4)}.$$

From this it follows that $\overline{AD} \cong \overline{BD}$.

PROPOSITION 11 *To draw a line at right angles to a given line from a given point on it.*

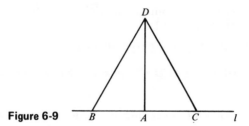

Figure 6-9

Construction Let l be the given line and A the given point (see Figure 6-9).

1. Determine on l the points B and C such that $\overline{AB} \cong \overline{AC}$ (Postulate III).
2. Construct an equilateral triangle BCD (Proposition 1).
3. Draw $\overline{DA}$ (Postulate I).

Then $\overline{DA}$ is the required perpendicular.

Proof

$$\left.\begin{array}{c} \overline{AB} \cong \overline{AC} \\ \overline{BD} \cong \overline{CD} \\ \overline{AD} \cong \overline{AD} \end{array}\right\} \quad \therefore \ \Delta ABD \cong \Delta ACD \quad \text{(Proposition 8)}.$$

From this it follows that $\angle BAD \cong \angle CAD$. Hence, $\overline{DA}$ is the perpendicular to l at A (Definition X).

PROPOSITION 12 *To drop a perpendicular to a given line from a given point not on the line.*

Construction Let l be the given line and A the given point (see Figure 6-10).

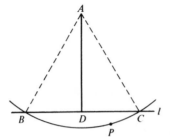

Figure 6-10

1. Choose a point P such that A and P lie on different sides of l.
2. Draw circle $(A, \overline{AP})$ (Postulate III). It cuts l in B and C.
3. Construct the midpoint, D, of $\overline{BC}$ (Proposition 10).
4. Draw $\overline{AD}$ (Postulate I).

Then $\overline{AD}$ is the required perpendicular.

Proof $\triangle ABD \cong \triangle ACD$ (Proposition 8). Hence, $\angle ADB$ is a right angle.

Remark   The four preceding constructions are existence proofs. In Proposition 11 the existence of the right angle is proved. Euclid did not define "different sides of a line" or "between" as is done today.

EXERCISES 6-7

1   Prove Proposition 5 without extending $\overline{CA}$ and $\overline{CB}$. (Hint: Prove $\triangle ABC \cong \triangle BAC$.)

2   Prove Proposition 6 but assume that $\overline{AC} < \overline{BC}$.

3   Prove Proposition 7 for each of the following cases:
   (a) D lies on side $\overline{AC}$.
   (b) D lies on side $\overline{AC}$ extended.
   (c) D lies inside $\triangle ABC$. (Hint: Use the second part of Proposition 5.)

4   Repeat the given proof of Proposition 8, by using a figure.

5   See the proof of Proposition 9. Change the proof such that $\triangle BCD$ is replaced by a triangle that is not equilateral.

6   See Figure 6-8, in which $\overline{AC} \cong \overline{BC}$ and $\overline{CD}$ bisects $\angle ACB$. Let P be a point of $\overline{CD}$ different from C and D. Prove that $\overline{PA} \cong \overline{PB}$.

7   Given a line segment $\overline{AB}$. Without extending $\overline{AB}$, construct a line segment $\overline{BC}$ such that $\overline{BC}$ is perpendicular to $\overline{AB}$.

8   Given a line segment $\overline{AB}$. Construct the line segments $\overline{AC}$ and $\overline{BD}$ such that both are perpendicular to $\overline{AB}$, $\overline{AC} \cong \overline{BD}$, and C and D are on opposite sides of line $\overleftrightarrow{AB}$. Prove that $\overline{AD} \cong \overline{BC}$.

9 Repeat Exercise 8 but assume that C and D are on the same side of line $\overleftrightarrow{AB}$.

10 See Exercise 8 and the corresponding figure. Draw $\overline{CD}$. Prove that $\angle ACD \cong \angle BDC$.

11 See Exercise 9 and the corresponding figure.
   (a) Prove that $\angle CAD \cong \angle DBC$.
   (b) Let P be the point of intersection of $\overline{AD}$ and $\overline{BC}$. Prove that $\overline{AP} \cong \overline{BP}$ and $\overline{CP} \cong \overline{DP}$.

6-8 CONGRUENCE (CONTINUED)

PROPOSITION 13 *If from a point of a line a ray is drawn, then this ray forms with the line two angles whose sum is congruent to the sum of two right angles.* (We shall sometimes use the name "ray" where Euclid uses the name "line.")

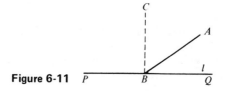

Figure 6-11

Proof Let l be the given line and $\overrightarrow{BA}$ the given ray (see Figure 6-11). If $\angle PBA \cong \angle ABQ$, then the two angles are right angles (Definition X). If $\angle PBA \not\cong \angle ABQ$, draw on the same side of l as $\overrightarrow{BA}$ a ray $\overrightarrow{BC} \perp l$ (Proposition 11). If $\overrightarrow{BC}$ lies inside $\angle PBA$, then we have

$$+ \begin{array}{l} \angle CBQ \quad\quad\quad \cong \angle CBA + \angle ABQ \\ \underline{\angle PBC \quad\quad\quad \cong \angle PBC} \\ \angle PBC + \angle CBQ \cong \angle PBC + \angle CBA + \angle ABQ \end{array} \quad \text{(Axiom II)}$$

and

$$+ \begin{array}{l} \angle PBA \quad\quad\quad \cong \angle PBC + \angle CBA \\ \underline{\angle ABQ \quad\quad\quad \cong \angle ABQ} \\ \angle PBA + \angle ABQ \cong \angle PBC + \angle CBA + \angle ABQ \end{array} \quad \text{(Axiom I).}$$

From this it follows that

$$\angle PBA + \angle ABQ \cong \angle PBC + \angle CBQ \quad \text{(Axiom I).}$$

Since $\angle PBC$ and $\angle CBQ$ are right angles, we have shown that the sum of $\angle PBA$ and $\angle ABQ$ is congruent to the sum of two right angles.

Remark The Greeks needed this proposition because they did not use the concept of a straight angle. In the following propositions the sum of two right angles will be represented by $2R$.

The converse of Proposition 13 is:

PROPOSITION 14   *If two angles have one side in common and if the noncommon sides are on different sides of the common side and if the angles are together congruent to $2R$, then the noncommon sides are the extensions of each other.*

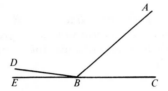

Figure 6-12

Proof   Assume that $\angle CBA + \angle ABD \cong 2R$ (see Figure 6-12). If $\overrightarrow{BD}$ is not the extension of $\overline{CB}$, then there is another ray, $\overrightarrow{BE}$, which is the extension of $\overline{CB}$ (Postulate II).

According to Proposition 13,

$$\angle CBA + \angle ABE \cong 2R,$$

and according to our assumption,

$$\angle CBA + \angle ABD \cong 2R.$$

Because of Postulate IV and Axiom II, the $2R$ in the first equation is equal to the $2R$ in the second equation, and hence, by Axiom I,

$$\angle CBA + \angle ABE \cong \angle CBA + \angle ABD,$$

from which it follows that

$$\angle ABE \cong \angle ABD \quad \text{(Axiom III)}.$$

However, this is contrary to Axiom VIII. Hence, $\overrightarrow{BD}$ is indeed the extension of $\overline{CB}$.

PROPOSITION 15   *Vertical angles are equal.*

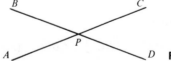

Figure 6-13

Proof   Consider the vertical angles *APB* and *CPD* (see Figure 6-13). According to Proposition 13,

$$\angle APB + \angle BPC \cong 2R$$

and $$\angle BPC + \angle CPD \cong 2R.$$

Then, by Postulate IV, Axiom II, and Axiom I,

$$\angle APB + \angle BPC \cong \angle BPC + \angle CPD.$$

From this it follows that

$$\angle APB \cong \angle CPD \quad \text{(Axiom III)}.$$

In the proofs of Propositions 1 through 15 the applied axioms, postulates, definitions, and propositions have been indicated. Since by this time the reader will have become familiar with the way in which Euclid supplied such references, we will sometimes omit them. It will improve the readability of the proofs.

PROPOSITION 16 *An exterior angle of a triangle is greater than each of the nonadjacent interior angles.*

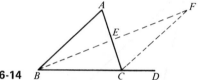

Figure 6-14

Proof Let $\angle ACD$ be an exterior angle of $\triangle ABC$ (see Figure 6-14). We shall prove that $\angle ACD > \angle A$.

Let E be the midpoint of $\overline{AC}$ (Proposition 10). Extend $\overline{BE}$ with a segment $\overline{EF}$ congruent to $\overline{BE}$. Draw $\overline{CF}$. Now we have

$$\begin{rcases} \overline{AE} \cong \overline{CE} \\ \overline{BE} \cong \overline{FE} \\ \angle AEB \cong \angle CEF \quad \text{(Proposition 15)} \end{rcases} \therefore \triangle ABE \cong \triangle CFE \\ \text{(Proposition 4)}.$$

From this it follows that

$$\angle ACF \cong \angle A.$$

By Axiom VIII,

$$\angle ACD > \angle ACF.$$

Hence, $$\angle ACD > \angle A.$$

Remarks

1. In the proof, Euclid concludes from Axiom VIII that $\angle ACD > \angle ACF$. For that purpose he tacitly assumes that point F is in the interior of $\angle ACD$. In a modern treatment this fact could be proved.

2. Euclid could have proved Proposition 16 with less effort if he could have used the theorem that in a triangle, the sum of the angles is congruent to $2R$. However, at the present stage this theorem has not been proved; it is a part of Proposition 32.

Why did Euclid not prove Proposition 32 before proving Proposition 16? By examining the proofs of Propositions 1 through 16, we observe that the first four postulates have been used frequently. Postulate V has not been used and its application will not occur until after Proposition 28. It looks as if Euclid tried to delay the application of Postulate V as long as possible. Proposition 32 is among those which cannot be proved without Postulate V. This may be the reason that Proposition 32 was not proved before Proposition 16.

PROPOSITION 17   *In a triangle, the sum of every two angles is less than 2R.*

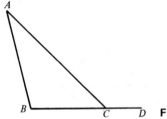

Figure 6-15

Proof   (See Figure 6-15.) We shall prove that $\angle B + \angle ACB < 2R$.
     Extend $\overline{BC}$. According to Proposition 16, $\angle ACD > \angle B$. Hence, we have

$$\angle B < \angle ACD$$
$$+ \ \underline{\ \ \ \angle ACB \cong \angle ACB \ \ \ }$$
$$\angle B + \angle ACB < \angle ACD + \angle ACB.$$

By Proposition 13,
$$\angle ACD + \angle ACB \cong 2R.$$

From this it follows that
$$\angle B + \angle ACB < 2R.$$

Remark   We note that in the proof the following property is used: If $\angle A < \angle B$, then $\angle A + \angle C < \angle B + \angle C$. Euclid did not include an axiom which justified this conclusion.

PROPOSITION 18   *If in a triangle one side is greater than another side, then the angle opposite the first side is greater than the angle opposite the other side.*

PROPOSITION 19   *If in a triangle one angle is greater than another angle, then the side opposite the first angle is greater than the side opposite the other angle.*

PROPOSITION 20   *In a triangle, the sum of two sides is greater than the third side.*

PROPOSITION 21   *If a point lies inside a triangle, then the sum of the connecting line segments of that point with the extremities of a side is less than the sum of the two other sides, and the angle at the point is greater than the angle opposite the side.*

We shall not give the proofs of these four propositions. The propositions themselves will not be needed to reach our announced goal: proving the Pythagorean theorem (see Exercises 2 through 5 of Exercises 6-8).

PROPOSITION 22   *To construct a triangle if the three sides are given.*

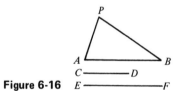

Figure 6-16

Construction   (See Figure 6-16.) Let $\overline{AB}$, $\overline{CD}$, and $\overline{EF}$ be the given line segments.

1. Draw circle $(A, \overline{CD})$.
2. Draw circle $(B, \overline{EF})$.
3. Let P be a point of intersection of the circles.

Then $\triangle ABP$ is the required triangle.

The construction is possible only if each of the three given sides is less than the sum of the other two (Proposition 20).

PROPOSITION 23   *To construct with a given ray as a side an angle that is congruent to a given angle.*

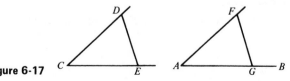

Figure 6-17

Construction   Let $\overrightarrow{AB}$ be the given ray and $\angle C$ the given angle (see Figure 6-17).

1. Choose on one side of $\angle C$ a point D and on the other side a point E.
2. Draw $\overline{DE}$.
3. Construct $\triangle AFG$ (G on $\overrightarrow{AB}$) such that $\overline{AF} \cong \overline{CD}$, $\overline{AG} \cong \overline{CE}$, and $\overline{GF} \cong \overline{ED}$ (Proposition 22).

Then $\angle FAG$ is the required angle.

Proof According to Proposition 8, $\triangle AFG \cong \triangle CDE$. Hence, $\angle A \cong \angle C$.

We shall omit Propositions 24 and 25. They concern two triangles having the property that two sides of one triangle are congruent to two sides of the other triangle and that the included angle of the first is not congruent to the included angle of the second.

PROPOSITION 26   *Two triangles are congruent if*
  (a) *one side and the two adjacent angles of one triangle are congruent to one side and the two adjacent angles of the other, or*
  (b) *one side, one adjacent angle, and the opposite angle of one triangle are congruent to one side, one adjacent angle, and the opposite angle of the other.*

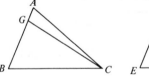 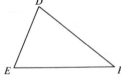

Figure 6-18

Proof of (a)   (See Figure 6-18.) Let $\overline{BC} \cong \overline{EF}$, $\angle B \cong \angle E$, and $\angle BCA \cong \angle F$. If

$$\overline{AB} \not\cong \overline{DE},$$

then one of $\overline{AB}$ and $\overline{DE}$ is the greater; assume that

$$\overline{AB} > \overline{DE}.$$

Choose G on $\overline{AB}$ such that

$$\overline{BG} \cong \overline{ED}.$$

Draw $\overline{GC}$. Then

$$\triangle GBC \cong \triangle DEF \quad \text{(Proposition 4)}.$$

Hence,

$$\angle BCG \cong \angle F.$$

Since it is given that

$$\angle F \cong \angle BCA,$$

it follows that

$$\angle BCG \cong \angle BCA,$$

which is contrary to Axiom VIII. Therefore,

$$\overline{AB} \cong \overline{DE},$$

and hence

$$\triangle ABC \cong \triangle DEF \quad \text{(Proposition 4)}.$$

Proof of (b)   (See again Figure 6-18.) Let $\overline{BC} \cong \overline{EF}$, $\angle B \cong \angle E$, and $\angle A \cong \angle D$. The proof is similar to that of (a). Again,

$$\triangle GBC \cong \triangle DEF.$$

From this it follows that

$$\angle BGC \cong \angle D$$

and hence that

$$\angle BGC \cong \angle A.$$

But this is contrary to Proposition 16. Hence, again, $\overline{AB} \cong \overline{DE}$ and $\triangle ABC \cong \triangle DEF$.

Remarks

1. The second part of this proposition would be an immediate consequence of the first part if we could use the theorem: The sum of the angles of a triangle is congruent to $2R$ (Proposition 32; see Remark 2 following Proposition 16).

2. Proposition 26(a) could be proved quite easily by means of a motion, in the same way as Proposition 4. Euclid does not do this but gives a far more complicated proof. *From this it appears that he objects to using a motion in a proof* and therefore avoids it if possible (see Remark 1 following Proposition 4).

3. The Propositions 4, 8, 26(a), and 26(b) are the familiar congruence theorems, which, in modern textbooks, are called side angle side (SAS), side side side (SSS), angle side angle (ASA), and side angle angle (SAA), respectively.

EXERCISES 6-8

1 See the proof of Proposition 16. Prove that $\angle ACD > \angle ABC$.

2 Prove Proposition 18.

3 Prove Proposition 19.

4 Prove Proposition 20.

5 Prove Proposition 21.

6 Of triangles ABC and $A'B'C'$ it is given that $\angle A \cong \angle A'$, $\overline{AC} \cong \overline{A'C'}$, $\overline{BC} \cong \overline{B'C'}$, and $\overline{BC} > \overline{AC}$ or $\overline{BC} \cong \overline{AC}$. Prove that the triangles are congruent.

7 See Exercise 6. Are the triangles necessarily congruent if $\overline{BC}$ is less than $\overline{AC}$?

8 Given: $\triangle ABC$, point D on $\overline{AC}$ between A and C, point E on $\overline{BC}$ between B and C. Prove: $\overline{AD} + \overline{DE} + \overline{EB} < \overline{AC} + \overline{CB}$.

9 Of triangles ABC and $A'B'C'$, $\overline{AB} \cong \overline{A'B'}$, $\overline{AC} \cong \overline{A'C'}$, and $\angle A > \angle A'$. Prove that $\overline{BC}$ is greater than $\overline{B'C'}$ (Proposition 24).

10 Of triangles ABC and $A'B'C'$, $\overline{AB} \cong \overline{A'B'}$, $\overline{AC} \cong \overline{A'C'}$, and $\overline{BC} > \overline{B'C'}$. Prove that $\angle A$ is greater than $\angle A'$ (Proposition 25).

11 Suppose that ABC is an equilateral triangle and D is a point of $\overline{BC}$ between B and C. Prove that $\overline{AD}$ is less than a side of $\triangle ABC$.

12 Suppose that ABC is an equilateral triangle, D is a point of $\overline{AB}$ between A and B, and E is a point of $\overline{BC}$ between B and C. Prove that $\overline{DE}$ is less than a side of $\triangle ABC$.

13 Suppose that ABC is an equilateral triangle, D is a point in its interior, and E is a point of $\overline{AB}$. Prove the following inequality:

$$\overline{AD} + \overline{DE} < \overline{AC} + \overline{CB}.$$

14 Is the inequality of Exercise 13 true for every triangle ABC with a point D in its interior and a point E on $\overline{AB}$?

15 Given: a line l and a point P not on l. Prove that the perpendicular segment from P to l is less than any other segment joining P and l.

6-9 THE THEORY OF PARALLELS

In Propositions 27 through 31 the theory of parallels is discussed.

PROPOSITION 27 *If two lines are intersected by a third line such that alternate interior angles are congruent, then the two lines are parallel.*

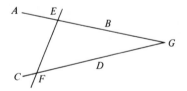

Figure 6-19

Proof Let it be given that $\angle AEF \cong \angle DFE$ (see Figure 6-19). Then we must prove that line $\overleftrightarrow{AB}$ is parallel to line $\overleftrightarrow{CD}$.

Assume that the lines $\overleftrightarrow{AB}$ and $\overleftrightarrow{CD}$ are not parallel. Then they meet in the direction of B and D or in the direction of A and C. Assume that they meet in the direction of B and D, and let G be the point of intersection. Then the exterior angle AEF of $\triangle EFG$ is greater than the interior angle GFE (Proposition 16). This is a contradiction of what is given. Hence, the lines $\overleftrightarrow{AB}$ and $\overleftrightarrow{CD}$ do not intersect in the direction of B and D.

Similarly, it can be proved that they do not intersect in the direction of A and C.

Hence, line $\overleftrightarrow{AB}$ is parallel to line $\overleftrightarrow{CD}$.

PROPOSITION 28 *If two lines are cut by a third line, then they are parallel if they satisfy one of the following conditions:*
(a) *Two corresponding angles are congruent.*
(b) *The sum of two interior angles on the same side of the transversal is congruent to 2R.*

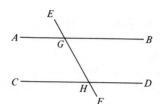

Figure 6-20

Proof of (a). (See Figure 6-20.) Let it be given that $\angle EGB \cong \angle EHD$. By Proposition 15 we have $\angle EGB \cong \angle AGF$. Hence, $\angle AGF \cong \angle EHD$. These two angles are alternate interior angles, and hence $\overleftrightarrow{AB} \parallel \overleftrightarrow{CD}$ (Proposition 27).

Proof of (b). (See again Figure 6-20.) Let it be given that $\angle FGB + \angle EHD \cong 2R$. By Proposition 13 we have $\angle FGB + \angle FGA \cong 2R$. Hence,

$\angle FGB + \angle FGA \cong \angle FGB + \angle EHD$ (Postulate IV, Axioms II
                                                   and I)

         $\angle FGA \cong \angle EHD$ (Axiom III)
            $\overleftrightarrow{AB} \parallel \overleftrightarrow{CD}$ (Proposition 27).

PROPOSITION 29 *If two parallel lines are cut by a third line, then:*
(a) *Alternate interior angles are congruent.*
(b) *Corresponding angles are congruent.*
(c) *The sum of two interior angles on the same side of the transversal is congruent to 2R.*

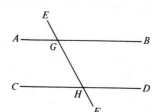

Figure 6-21

Proof of (a).   (See Figure 6-21.) Let it be given that $\overleftrightarrow{AB}$ and $\overleftrightarrow{CD}$ are the parallel lines and that $\overleftrightarrow{EF}$ is the third line. We shall prove that the alternate interior angles AGF and DHE are congruent.

If they are not congruent, then one of them is the smaller. Let $\angle DHE$ be the smaller. Then we have

$$\angle DHE < \angle AGF$$
$$+ \quad \underline{\angle BGF \cong \angle BGF}$$
$$\angle DHE + \angle BGF < \angle AGF + \angle BGF$$

or, by Proposition 13,

$$\angle DHE + \angle BGF < 2R.$$

Hence, by Postulate V, $\overleftrightarrow{AB}$ and $\overleftrightarrow{CD}$ have a point of intersection, contrary to what was given. From this it follows that the angles AGF and DHE are congruent.

We leave the proofs of (b) and (c) to the reader. (See Exercises 3 and 4 of Exercises 6-9.)

Remark   We notice that in the preceding proof Postulate V has been used. Proposition 29a (and 29b and 29c) could not have been proved without it. This is the first time that Euclid made use of Postulate V. Because of its relation to the concept of parallelism, Postulate V is often called the *Parallel Postulate* or *Parallel Axiom*.

PROPOSITION 30   *Lines that are parallel to the same line are parallel to each other.*

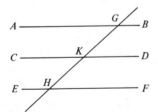

Figure 6-22

Proof   (See Figure 6-22.) Let it be given that $\overleftrightarrow{AB} \parallel \overleftrightarrow{EF}$ and $\overleftrightarrow{CD} \parallel \overleftrightarrow{EF}$. We shall prove that $\overleftrightarrow{AB} \parallel \overleftrightarrow{CD}$.

Draw the transversal $\overleftrightarrow{GK}$, where G is on $\overleftrightarrow{AB}$ and K on $\overleftrightarrow{CD}$. Then we have

$$\angle AGH \cong \angle FHG \quad \text{(Proposition 29a)}$$
$$\angle FHG \cong \angle DKG \quad \text{(Proposition 29b)}.$$

From this it follows that

$$\angle AGH \cong \angle DKG \quad \text{(Axiom I)}$$

and hence that

$$\overleftrightarrow{AB} \parallel \overleftrightarrow{CD} \quad \text{(Proposition 27)}.$$

Remarks

1. We pointed out before that it was evidently Euclid's intention to deduce nothing from a figure, but that he was not always successful. The proof of Proposition 30 contains another example of a conclusion that Euclid deduced from a figure. In this proof, Euclid assumes that if a line ($\overleftrightarrow{GK}$) cuts one of two parallel lines ($\overleftrightarrow{AB}$), then it also cuts the other ($\overleftrightarrow{EF}$). However, this is not a theorem that has been proved previously.

2. We shall state and prove the theorem referred to in Remark 1.

PROPOSITION A *If a line cuts one of two parallel lines, then it also cuts the other.*

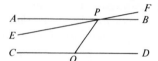

Figure 6-23

The proof is as follows (see Figure 6-23).

Let $\overleftrightarrow{AB}$ be parallel to $\overleftrightarrow{CD}$ and let $\overleftrightarrow{EF}$ be a line that cuts $\overleftrightarrow{AB}$ in P. We shall prove that $\overleftrightarrow{EF}$ intersects $\overleftrightarrow{CD}$.

Assume that $\overleftrightarrow{EF} \parallel \overleftrightarrow{CD}$. Choose on $\overleftrightarrow{CD}$ a point Q and draw $\overleftrightarrow{PQ}$. Since $\overleftrightarrow{AB} \parallel \overleftrightarrow{CD}$,

$$\angle APQ \cong \angle DQP \quad \text{(Proposition 29a)}.$$

Since $\overleftrightarrow{EF} \parallel \overleftrightarrow{CD}$,

$$\angle EPQ \cong \angle DQP \quad \text{(Proposition 29a)}.$$

Hence, $\angle APQ \cong \angle EPQ,$

which is contrary to Axiom VIII. Hence, $\overleftrightarrow{EF}$ is not parallel to $\overleftrightarrow{CD}$; that is, $\overleftrightarrow{EF}$ intersects $\overleftrightarrow{CD}$.

3. From Proposition A it follows (see Figure 6-23) that no two lines, both parallel to $\overleftrightarrow{CD}$, can go through P. Hence, we have:

PROPOSITION B *Through a given point not on a given line, not more than one line can be drawn parallel to the given line.*

4. In proving Propositions A and B, we have made use of Proposition 29a, which is based on Postulate V. Therefore, Propositions A and B are consequences of Postulate V.

5. The previous remarks show that we have the following connections between Postulate V and Propositions 29a, A, and B, where the symbol " ⇒" means "implies":

$$V \Rightarrow 29a \Rightarrow A \Rightarrow B. \tag{1}$$

It turns out that we also have the following connections:

$$B \Rightarrow 29a \Rightarrow 29c \Rightarrow V \tag{2}$$

(see Problem 5 of Exercises 6-9).

Propositions 1 through 28 have been proved without making use of Postulate V. In chain (1), we prove Proposition B by using Propositions 1 through 28 and by assuming that Postulate V is true (together with Postulates I through IV). In chain (2), we prove Postulate V by using Propositions 1 through 28 and by assuming that Proposition B is true (together with Postulates I through IV). In (1) and (2) we end up with the same geometrical structure.

From this discussion we see that it is sometimes possible to build up the same geometrical system starting from different sets of assumptions.

6. In Remark 5 we showed that the geometrical structure is not changed when the Parallel Postulate is replaced by Proposition B. Since Proposition B is easier to understand than Postulate V, in modern books Proposition B is usually given in place of Euclid's Parallel Postulate.

7. Generations of mathematicians have attempted to prove Postulate V. The reason for these attempts may have been that the converse of Postulate V could be proved; it is Proposition 27. This is the only case in Book I of the *Elements* where a theorem was true but its converse could not be proved. In the eighteenth and nineteenth centuries mathematicians began to realize that Postulate V probably could not be proved, but that if this postulate was altered, a different geometry could be developed, one that was consistent with the new postulate. *Saccheri's* false proof of Postulate V contained the first *non-Euclidean geometry.* Independent from Saccheri and from each other, *Gauss, Lobatchevski,* and *Bolyai* developed a non-Euclidean geometry by assuming, instead of the Parallel Postulate, the equivalent of: Through a given point not on a given line, there is more than one line through the point parallel to the given line. This choice of a postulate in defiance of common sense profoundly affected the attitude of later mathematicians toward postulates. Postulates were no longer perceived as obvious truths but were considered to be *assumptions* upon which a mathematical structure was based.

PROPOSITION 31 *Given a line and a point not on the line, to construct through the point a line parallel to the given line.*

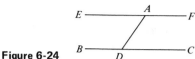

Figure 6-24

Construction Let A be the given point and $\overleftrightarrow{BC}$ the given line (see Figure 6-24).

1. Choose a point D on $\overleftrightarrow{BC}$.
2. Draw $\overleftrightarrow{AD}$.
3. Draw $\overleftrightarrow{EF}$ through A such that $\angle EAD \cong \angle CDA$.

Then $\overleftrightarrow{EF}$ is the required line.

Proof From $\angle EAD \cong \angle CDA$ it follows that $\overleftrightarrow{EF} \parallel \overleftrightarrow{BC}$ (Proposition 27).

Remarks

1. Proposition B in combination with Proposition 31 could be formulated as follows: Through a given point not on a given line there is exactly one line parallel to the given line. This statement is sometimes called Playfair's postulate, after *John Playfair* (1748-1819).

2. The construction proves the existence of parallel lines, in agreement with Aristotle's requirements.

PROPOSITION 32 (a) *An exterior angle of a triangle is congruent to the sum of the nonadjacent interior angles.*
(b) *The sum of the angles of a triangle is congruent to 2R.*

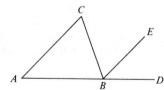

Figure 6-25

Proof of (a) Let ABC be the triangle (see Figure 6-25). Extend $\overrightarrow{AB}$ to D. Draw $\overrightarrow{BE} \parallel \overline{AC}$ (E on the same side of $\overleftrightarrow{AB}$ as C). Then we have

$$\begin{aligned}
\angle CBE &\cong \angle C \quad \text{(Proposition 29a)} \\
+ \quad \underline{\angle DBE \cong \angle A} \quad &\text{(Proposition 29b)} \\
\angle CBD &\cong \angle A + \angle C.
\end{aligned}$$

Proof of (b). (See again Figure 6-25.) Since

$$\angle CBD + \angle CBA \cong 2R,$$

we also have

$$\angle A + \angle C + \angle CBA \cong 2R.$$

Hence, in every triangle *ABC,*

$$\angle A + \angle B + \angle C \cong 2R.$$

PROPOSITION 33   *If two opposite sides of a quadrilateral are congruent and parallel, then the other two sides are also congruent and parallel.*

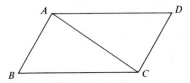

Figure 6-26

Proof   Let $\overline{AD} \cong \overline{BC}$ and $\overline{AD} \parallel \overline{BC}$ (see Figure 6-26). Draw $\overline{AC}$. Then

$\left.\begin{array}{l} \angle BCA \cong \angle DAC \quad \text{(Proposition 29a)} \\ \overline{AC} \cong \overline{CA} \\ \overline{BC} \cong \overline{DA} \end{array}\right\} \therefore \Delta ABC \cong \Delta CDA \quad \text{(SAS)}.$

Hence, $\overline{AB} \cong \overline{CD}$ and $\angle BAC \cong \angle DCA$, from which it follows that $\overline{AB} \parallel \overline{DC}$ (Proposition 27).

PROPOSITION 34.   *In a parallelogram:*
  (a) *The opposite sides are congruent.*
  (b) *The opposite angles are congruent.*

The proof is left to the reader (see Exercise 7 of Exercises 6-9; Euclid does not give an explicit definition of a parallelogram, but from his proof of Proposition 34, it appears that he uses the following definition: A parallelogram is a quadrilateral with two pairs of parallel sides).

EXERCISES 6-9

1 Prove the theorem: If two lines are cut by a third line, then they are parallel if alternate exterior angles are congruent,
   (a) by applying Proposition 28a,
   (b) by applying Proposition 28b.

2 Prove the theorem: If two lines are cut by a third line, then they are parallel if the sum of two exterior angles on the same side of the transversal is congruent to $2R$,

(a) by applying Proposition 28a,

(b) by applying Proposition 28b.

3 Prove Proposition 29b.

4 Prove Proposition 29c.

5 See chain (2) on page 172. Prove:
(a) Proposition B ⇒ Proposition 29a,
(b) Proposition 29a ⇒ Proposition 29c,
(c) Proposition 29c ⇒ Postulate V.

6 See Remark 2 following Proposition 31. Could Euclid have proved the existence of parallel lines before proving the propositions that are based on Postulate V?

7 Prove Propositions 34a and 34b.

8 Prove that the sum of the angles of a quadrilateral is congruent to 4R.

9 Prove that the diagonals of a parallelogram bisect each other.

10 Prove that the diagonals of a rhombus are perpendicular to each other.

11 Prove: If the diagonals of a quadrilateral bisect each other, then the quadrilateral is a parallelogram.

12 Prove: If the diagonals of a quadrilateral bisect each other and are perpendicular to each other, then the quadrilateral is a rhombus.

13 Disprove: If the angles A and C of quadrilateral $ABCD$ are congruent, then $ABCD$ is a parallelogram.

14 Prove that the base angles of an isosceles trapezoid are congruent.

15 Prove: If the base angles of a trapezoid are congruent, then the trapezoid is isosceles.

16 Prove: If the diagonals $\overline{AC}$ and $\overline{BD}$ of trapezoid $ABCD$ are congruent, then $ABCD$ is isosceles. (Hint: Draw a line through C parallel to $\overline{BD}$.)

17 Prove that the line segment joining the midpoints D and E of the sides $\overline{AC}$ and $\overline{BC}$ of triangle ABC is parallel to $\overline{AB}$. (Hint: Extend $\overline{DE}$ with a segment $\overline{EF}$ congruent to $\overline{DE}$.)

6-10 THE COMPARISON OF AREAS

The Propositions 34c through 41, most of which are needed for the proof of the theorem of Pythagoras, deal with the comparison of areas.

PROPOSITION 34c *A parallelogram is divided by a diagonal into two triangles equal in area.*

> *Proof* We have already seen that a parallelogram is divided by a diagonal into two congruent triangles. (See Exercise 7 of Exercises 6-9.) From the congruence of the triangles, it follows that they are equal in area (Axiom VII).

PROPOSITION 35   *Two parallelograms with the same base and lying between the same parallel lines are equal in area.*

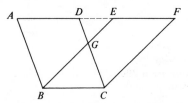

Figure 6-27

Proof   (See Figure 6-27.) Let *ABCD* and *EBCF* be the parallelograms and let $\overline{AD}$ and $\overline{EF}$ lie on one straight line parallel to $\overline{BC}$. Then $\overline{AD} \cong \overline{BC}$ (Proposition 34a), and likewise $\overline{EF} \cong \overline{BC}$. Hence, $\overline{AD} \cong \overline{EF}$. From this it follows that $\overline{AE} \cong \overline{DF}$. We now have

$\overline{AE} \cong \overline{DF}$

$\angle A \cong \angle CDF$ (Proposition 29b) $\Bigg\}$ $\therefore \triangle ABE \cong \triangle DCF$ (SAS).

$\overline{AB} \cong \overline{DC}$ (Proposition 34a)

Hence,

$$\triangle ABE = \triangle DCF \quad \text{(Axiom VII)}$$
$$- \ \frac{\triangle DGE = \triangle DGE}{ABGD = EGCF}$$
$$+ \ \frac{\triangle GBC = \triangle GBC}{ABCD = EBCF.}$$

Remarks

1. In the previous proof we notice that Euclid speaks of equal figures, where "equal" does not mean "can be made to coincide." (See Remark 1 following Proposition 1.) Here, Euclid must have had an intuitive idea of the amount of space (area) enclosed by the boundary of a figure. He does not attach a numerical value to this amount of space.

2. The proof depends on Figure 6-27. It does not cover the case that *E* is between *A* and *D*. (See Exercise 1 of Exercises 6-10.)

PROPOSITION 36   *Two parallelograms with congruent bases and lying between the same parallel lines are equal in area.*

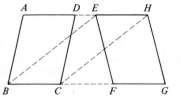

Figure 6-28

Proof (See Figure 6-28.) Let *ABCD* and *EFGH* be the parallelograms. Then $\overleftrightarrow{AD}$ and $\overleftrightarrow{EH}$ lie on one line, $\overleftrightarrow{BC}$ and $\overleftrightarrow{FG}$ lie on another line, and these lines are parallel. Further, $\overline{BC} \cong \overline{FG}$. We shall prove that *ABCD* = *EFGH*.

Draw $\overline{BE}$ and $\overline{CH}$. Then,

$$\left.\begin{array}{l} \overline{FG} \cong \overline{EH} \quad \text{(Proposition 34a)} \\ \overline{BC} \cong \overline{FG} \end{array}\right\} \quad \therefore \ \overline{BC} \cong \overline{EH}.$$

Since $\overline{BC} \parallel \overline{EH}$, it follows that *EBCH* is a parallelogram (Proposition 33). Hence,

$$\left.\begin{array}{l} EBCH = ABCD \quad \text{(Proposition 35)} \\ EBCH = EFGH \quad \text{(Proposition 35)} \end{array}\right\} \quad \therefore \ ABCD = EFGH.$$

PROPOSITION 37 *Two triangles with the same base and lying between the same parallel lines are equal in area.*

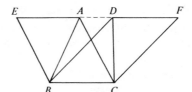

Figure 6-29

Proof (See Figure 6-29.) Let *ABC* and *DBC* be the triangles and $\overleftrightarrow{AD} \parallel \overleftrightarrow{BC}$.

Draw $\overline{BE} \parallel \overline{CA}$ and $\overline{CF} \parallel \overline{BD}$ (Proposition 31). Then *EBCA* and *DBCF* are parallelograms. By Proposition 35 they are equal in area. Since $\overline{AB}$ and $\overline{CD}$ are diagonals,

$$\left.\begin{array}{l} EBCA = 2(\triangle ABC) \quad \text{(Proposition 34c)} \\ DBCF = 2(\triangle DBC) \quad \text{(Proposition 34c)} \end{array}\right\} \quad \therefore \ \triangle ABC = \triangle DBC.$$

PROPOSITION 38 *Two triangles with congruent bases and lying between the same parallel lines are equal in area.*

Proof The proof is analogous to that of Proposition 37 and is based on Proposition 36. (See Exercise 2 of Exercises 6-10.)

Propositions 39 and 40 are converses of Propositions 37 and 38. They will not be discussed here. (See Exercises 3 and 4 of Exercises 6-10.)

PROPOSITION 41 *If a parallelogram and a triangle have the same base and lie between the same parallel lines, then the parallelogram is double the triangle.*

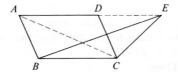

Figure 6-30

Proof   (See Figure 6-30.) Let $\underleftrightarrow{ABCD}$ be the parallelogram and EBC the triangle, where E lies on $\overleftrightarrow{AD}$. Draw $\overline{AC}$. Then

$$\left. \begin{array}{ll} \triangle ABC = \triangle EBC & \text{(Proposition 37)} \\ ABCD = 2(\triangle ABC) & \text{(Proposition 34c)} \end{array} \right\} \quad \therefore ABCD = 2(\triangle EBC).$$

EXERCISES 6-10

1 See Proposition 35 and the accompanying illustration, Figure 6-27. Prove the proposition for the case that point E is between A and D.

2 Give the details of the proof of Proposition 38.

3 Prove: **Triangles with the same area and the same base and which are on the same side of the base lie between the same parallel lines** (Proposition 39).

4 Of two triangles the following is given:
   (1) They are equal in area.
   (2) Their bases are congruent.
   (3) The bases are segments of the same line.
   (4) The triangles lie on the same side of that line.
   Prove that the triangles lie between the same parallel lines (Proposition 40).

5 Prove that a median of a triangle separates the triangle into two parts equal in area.

6 In $\triangle ABC$, D is the midpoint of $\overline{AC}$ and E is the midpoint of $\overline{BC}$. Prove that quadrilateral $ABED$ is three times $\triangle DEC$. (Hint: See Exercise 17 of Exercises 6-9.)

7 Construct a triangle that is three times a given triangle.

8 In a parallelogram $ABCD$, the following is given: P is a point of diagonal $\overline{BD}$; $\overline{EF}$ passes through P and is parallel to $\overline{AB}$ (E on $\overline{AD}$, F on $\overline{BC}$); $\overline{GH}$ passes through P and is parallel to $\overline{AD}$ (G on $\overline{AB}$, H on $\overline{DC}$). Prove: $AGPE = PFCH$ (Proposition 43).

9 Given: a parallelogram $ABCD$ and a line segment $\overline{PQ}$. Construct on $\overline{PQ}$ a parallelogram $PQRS$ equal in area to $ABCD$. (Hint: See Exercise 8.)

10 Given: a parallelogram $ABCD$, a line segment $\overline{PQ}$, and an angle KLM. Construct on segment $\overline{PQ}$ a parallelogram $PQRS$ equal in area to $ABCD$ and such that $\angle SPQ \cong \angle KLM$.

6-11   THE THEOREM OF PYTHAGORAS

The other theorems concerning comparison of areas (Propositions 42 through 45) are not needed for our final goal, the proof of the theorem of Pythagoras. The Greeks formulated this theorem somewhat differently from the way we do. We speak of the squares of the sides of a right triangle (that is, the squares of the numbers that represent the lengths of those sides). The Greeks, on the other hand, spoke of the squares constructed on the sides of the triangle.

Therefore, since the Pythagorean theorem speaks of geometrical squares, we will first give the Euclidean definition of a square. It is as follows: *A square is a quadrilateral all sides of which are congruent and all angles of which are right angles.*

Before proving a theorem about squares, we must prove that squares exist. This is the purpose of Proposition 46.

PROPOSITION 46   *To construct a square on a given line segment.*

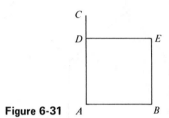

Figure 6-31

Construction   Let $\overline{AB}$ be the given line segment (see Figure 6-31).
1. Draw $\overline{AC} \perp \overline{AB}$ (Proposition 11).
2. Determine a point D on $\overline{AC}$ such that $\overline{AD} \cong \overline{AB}$.
3. Draw through D a line parallel to $\overline{AB}$ (Proposition 31).
4. Draw through B a line parallel to $\overline{AD}$ (Proposition 31).
5. Let E be the point of intersection of these lines.

Then *ABED* is the required square.

Proof   We first prove that the sides are congruent.

$\overline{BE} \parallel \overline{AD}$ (by construction) $\Big\}$ $\therefore \ \overline{AD} \cong \overline{BE}$ and $\overline{AB} \cong \overline{DE}$
$\overline{DE} \parallel \overline{AB}$ (by construction) $\Big\}$ (Proposition 34a) .

Moreover,

$$\overline{AD} \cong \overline{AB} \quad \text{(by construction)}.$$

Hence, $\overline{AB} \cong \overline{BE} \cong \overline{ED} \cong \overline{DA}.$

Next we prove that the angles are right angles. Since $\overline{DE} \parallel \overline{AB}$, we have

$$\angle A + \angle ADE \cong 2R \quad \text{(Proposition 29c)}.$$

$\angle A$ is a right angle (thus constructed). Hence, $\angle ADE$ is also a right angle.

$$\angle ADE \cong \angle B \quad \text{and} \quad \angle A \cong \angle E \quad \text{(Proposition 34b).}$$

Hence,

$$\angle A, \ \angle B, \ \angle E, \ \text{and} \ \angle ADE \ \text{are right angles.}$$

From this it follows that $ABED$ is a square.

PROPOSITION 47   *"Theorem of Pythagoras." In a right triangle, the square on the hypotenuse is equal in area to the sum of the squares on the sides.*

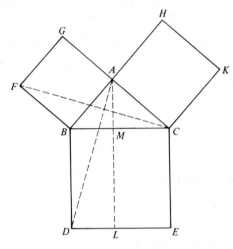

Figure 6-32

Proof   (See Figure 6-32.) Let ABC be a triangle in which $\angle A$ is a right angle. Construct the squares $AGFB$, $BDEC$, and $CKHA$ (Proposition 46). Draw through A a line parallel to $\overline{BD}$ (Proposition 31); it cuts $\overline{DE}$ in L and $\overline{BC}$ in M. Draw $\overline{AD}$ and $\overline{CF}$. Then we have

$$\angle BAC + \angle BAG \cong 2R.$$

Hence, $\overline{AG}$ is the extension of $\overline{CA}$. We now have

$$\begin{array}{l} \angle DBC \cong \angle ABF \\ + \ \dfrac{\angle CBA \cong \angle CBA}{\left.\begin{array}{l} \angle DBA \cong \angle CBF \\ \overline{BA} \cong \overline{BF} \\ \overline{BD} \cong \overline{BC} \end{array}\right\}} \end{array} \quad \therefore \ \triangle DBA \cong \triangle CBF \quad \text{(SAS).}$$

We also have

$$DBML = 2(\triangle DBA) \quad \text{(Proposition 41)}$$

$$ABFG = 2(\triangle CBF) \quad \text{(Proposition 41).}$$

Hence, $DBML = ABFG.$ (1)

In a similar way we can prove that

$$ECML = ACKH.$$ (2)

From statements (1) and (2) it follows that

$$DBCE = ABFG + ACKH.$$

EXERCISES 6-11

1 See the proof of Proposition 47. Prove statement (2), that $ECML = ACKH.$

2 Prove the converse of Proposition 47: If the sum of the squares on two sides of a triangle is equal to the square on the third side, then the triangle is a right triangle (Proposition 48, the last proposition of Book I of the *Elements*).

3 See Figure 6-32. Prove that $\overline{AD}$ is perpendicular to $\overline{CF}$.

4 See Figure 6-32. Let S be the point of intersection of $\overleftrightarrow{FG}$ and $\overleftrightarrow{KH}$. Draw $\overline{AS}$.
   (a) Prove that $\triangle AHS \cong \triangle CAB.$
   (b) Prove that the points S, A, and M are on one line.

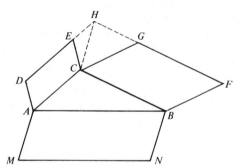

Figure 6-33

5 The theorem contained in this problem is due to Pappus and is a generalization of the theorem of Pythagoras. It shows how generalizations of some known mathematics can lead to interesting new mathematics.

   In Figure 6-33, the quadrilaterals $ACED$, $CBFG$, and $ABNM$ are parallelograms. $\overleftrightarrow{DE}$ and $\overleftrightarrow{FG}$ intersect in the point H. $\overline{AM}$ is parallel and congruent to $\overline{HC}$. Prove that $ACED + CBFG = ABNM$. (Hint: Extend $\overline{HC}$, $\overline{MA}$, and $\overline{NB}$ until they meet $\overline{MN}$, $\overline{DE}$, and $\overline{FG}$, respectively).

6 Construct a parallelogram equal in area to the sum of two given parallelograms. (Hint: See Exercise 5.)

7 Prove Proposition 47 by using the method of Exercise 5.

8 See the construction in Proposition 46. Prove that point E exists.

6-12   THE DIFFERENCE BETWEEN THE EUCLIDEAN AND
THE MODERN METHOD OF COMPARING AREAS

We note that Euclid's way of comparing areas (from Proposition 34 on)
is quite different from contemporary treatments. In modern treatments
we indicate the area of a figure (just as the length of a line segment) by
a number. We also state formulas for the areas of different figures. We
thus find, for example, that the area of a parallelogram is equal to $b \cdot h$.
That is, the number that indicates the area of the parallelogram in
square units is equal to the product of the two numbers that indicate
the lengths of the base and the altitude in units of length.

*Euclid, however, indicates neither the lengths of line segments nor
the areas of figures by numbers. If he wants to show that two figures
have equal areas, he can do so only by demonstrating that one figure
can be divided into parts, such that these parts, if fitted together in a
certain way, produce the other figure.* (In fact, his process is even more
complicated, but this does not affect the general idea.)

To illustrate clearly the difference between these two methods, we
choose as an example the modern proof of Proposition 35: *Two parallelograms with the same base and lying between the same parallel lines
are equal in area.* This proof is as follows. Since the parallelograms lie
between the same parallel lines, the altitudes are equal in length. Also
the bases are equal in length. Hence, the product $b \cdot h$ is the same for
the two parallelograms, and hence their areas are equal. Euclid does
not have the proposition that the area of a parallelogram is equal to the
product of the length of the base and the length of the altitude. He is
therefore obliged to give another proof. Compare the proof given on
page 176 with the one given here.

Why did Euclid not indicate the length of a line segment and the
area of a figure by a number? For an answer to this question we again
recall the development of Greek thought. The Pythagoreans discovered
that the natural numbers play a part outside mathematics (music!).
This led them to the practice of attaching mystical power to the numbers. Everything could be reduced to number. *However, their numbers
were exclusively natural numbers.* Also with Plato the natural number
plays a fundamental part, but on quite different grounds. For him the
unit is an idea, a basic philosophical concept. In the material world one
finds things that are considered as a unit (remind us of the unit). To
quantities of these things, a number is assigned. But then that number
is always a natural number, for the absolute unit is indivisible. *The use
of numbers other than natural numbers is forbidden by Plato for
philosophical reasons.*

Let us suppose for a moment that the Greeks had taken a line segment as a unit of measurement and compared all other line segments

with it. Call this line segment e. Then to determine the length of another line segment, a, simply means to determine the ratio of a to the chosen unit of length e. Hence, if the ratio of the line segment a to the unit of length e is rational, no difficulties arise. We then say that the length of the line segment a is, for instance, equal to $\frac{3}{4}$. The Greeks would say that the line segments a and e are proportional to the numbers 3 and 4. There is no essential difficulty here. *However, as soon as the ratio of the line segments is an irrational number, they got into insurmountable difficulties.* When the length of the line segment a is equal to $\sqrt{2}$, for instance, there are no two natural numbers that are to each other as the lengths of a and e. Hence, it was impossible for the Greeks to indicate the length of such a line segment by means of numbers. So, for some line segments they could have represented the length by means of numbers; for others, they could not. It is understandable that for this reason they abandoned all attempts to express lengths of line segments in terms of numbers, even in the case where the ratio between the line segment and the unit could have been expressed by using natural numbers. Not only was it impossible for the Greeks to denote the length of each line segment by a number, but it was also impossible for them to express each area by a number. Therefore, their theory of areas, which consisted exclusively of a comparison of areas, did not make use of numbers.

It is now clear why the Greeks formulated the Pythagorean theorem in a different way than we do. A typical modern formulation is the following: If $\angle C$ of $\triangle ABC$ is a right angle, then $a^2 + b^2 = c^2$. In this formula, a, b, and c are numbers, the lengths of the sides of $\triangle ABC$. Since they did not use numbers for lengths, the Greeks of necessity formulated the theorem differently. Euclid's formulation was: In a right triangle, the square on the hypotenuse is equal to the sum of the squares on the sides. The proof had to be given by comparison of areas.

In the next two sections we shall discuss some topics from other books of the *Elements*. Although Euclid wrote all the books of the *Elements* in the rigorous manner that we have seen in Book I, we shall omit many propositions from now on and, consequently, adopt a more informal style of writing.

6-13 GEOMETRIC ALGEBRA AND REGULAR POLYGONS

Today, the first propositions of Book II of the *Elements* would be considered algebraic identities. Proposition II, 1 (which means Proposition 1 of Book II) is as follows: *If there be two straight lines, one of which is divided into any number of parts, the rectangle contained by the two*

straight lines is equal in area to the rectangles contained by the un-
divided line and the several parts of the divided line (see Figure 6-34).

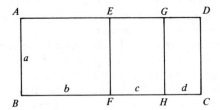

Figure 6-34

A modern formulation of the proposition is the following: If in rectangle *ABCD* one side has the length *a* and the other side consists of
three parts with lengths *b*, *c*, and *d*, then $a(b + c + d) = ab + ac + ad$.
This identity represents the distributive property of multiplication over
addition. We may regard the variables *a*, *b*, *c*, and *d* as symbols that can
be replaced by real numbers. (However, as we have seen in the previous
section, Euclid did not associate with a line segment a number called its
length or with a rectangle a number called its area.) Other examples of
this type of geometric algebra are to be found in Exercises 1 and 2 of
Exercises 6-13.

Proposition II, 11 involves a type of geometric algebra related to
the solution of quadratic equations with irrational roots. The proposition is as follows: *To divide a given straight line into two parts so
that the rectangle contained by the whole and one of the parts is equal
in area to the square on the other part.* That is, if $\overline{AB}$ is a given line segment (see Figure 6-35), we must construct a point *X* on $\overline{AB}$ such that
the rectangle with sides of lengths *AB* and *XB* has the same area as the
square on $\overline{AX}$.

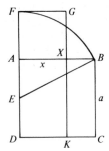

Figure 6-35

Let $AB = a$ and $AX = x$. Then the construction amounts to solving the equation

$$a(a - x) = x^2,$$

or
$$x^2 + ax - a^2 = 0, \tag{1}$$

a quadratic equation with the roots

$$\tfrac{1}{2}a\left(-1+\sqrt{5}\right) \text{ and } \tfrac{1}{2}a\left(-1-\sqrt{5}\right).$$

Euclid's solution of the proposition amounts to constructing the first of these roots. (The second root is negative.)

Construct the square $ADCB$ and the midpoint E of $\overline{AD}$. Construct the point of intersection F of circle $(E, \overline{EB})$ and the extension of $\overline{DA}$. Construct square $AXGF$. Then X is the required point.

Euclid's proof is as follows (in simplified form and by use of modern notation):

$$AE = \tfrac{1}{2}AD = \tfrac{1}{2}a$$

Hence, in the right triangle EAB we have

$$EB = \sqrt{a^2 + (\tfrac{1}{2}a)^2} = \tfrac{1}{2}a\sqrt{5}.$$

Hence,
$$AF = EF - EA$$
$$= EB - EA$$
$$= \tfrac{1}{2}a\sqrt{5} - \tfrac{1}{2}a = \tfrac{1}{2}a(-1+\sqrt{5}).$$

Equation (1) is a special form of a quadratic equation. Later books of the *Elements* contain geometric constructions that are equivalent to solving a general quadratic equation.

In Proposition IV, 11 Euclid gives a construction of the regular pentagon. For that purpose he first shows (in Proposition IV, 10) how *to construct an isosceles triangle with each base angle equal (in measure) to two times the vertex angle.* We shall give a proof of Proposition IV, 10 by use of modern notation and angle measure. (Angle measure was never used by Euclid; see Remark 1 following Proposition I, 1.)

Suppose that ABC is such a triangle, with $AC = BC$ (see Figure 6-36). Then the measures of angles A, B, and C are $72°$, $72°$, and $36°$, respectively.

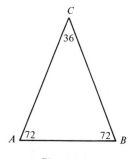

Figure 6-36

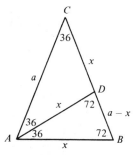

Figure 6-37

Let $\overline{AD}$ be the bisector of $\angle A$ (see Figure 6-37). Then

$$m\angle BAD = 36° = m\angle C,$$

from which it follows that $\triangle ABC \sim \triangle DBA$ (congruent angles).

Hence, $AB = AD.$

Since $m\angle CAD = 36° = m\angle C,$

we have $CD = AD.$

Let $AC = a$ and $AB = x$. Then

$$AB = AD = CD = x$$

and $BD = BC - CD = a - x.$

From the similarity of triangles ABC and DBA it also follows that

$$\frac{AB}{CB} = \frac{DB}{AB}$$

and hence that $\dfrac{x}{a} = \dfrac{a-x}{x}.$

This proportion leads to equation (1),

$$x^2 + ax - a^2 = 0.$$

The positive root,

$$\tfrac{1}{2}a\left(-1 + \sqrt{5}\right),$$

can be constructed if a line segment of length a is given (see Figure 6-35). From this it follows that we can construct a triangle ABC with $AC = BC$ and $m\angle C = 36°$, that is, a triangle as required in Proposition IV, 10.

For the purpose of constructing a regular pentagon we observe the following. If a regular pentagon $ABCDE$ is inscribed in a circle (see Figure 6-38), then the measure of each of the arcs in which the circle is divided is $360°/5 = 72°$. Draw AD and BD. Then $AD = BD$ (chords of congruent arcs) and $m\angle ADB = \tfrac{1}{2}m$ (arc AB) $= 36°$.

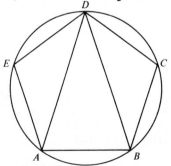

Figure 6-38

Hence, the regular pentagon can be constructed as follows:

1. Take an arbitrary line segment; let a be its length.
2. Construct a line segment of length $x = \frac{1}{2}a(-1 + \sqrt{5})$.
3. Construct the isosceles triangle ABD with sides of lengths x, a, and a.
4. Circumscribe a circle about this triangle.
5. Complete the pentagon.

In our construction of the regular pentagon $ABCDE$ we first chose a diagonal $\overline{AD}$ of length a; then we constructed $\triangle ABD$ with angles of $36°$, $72°$, and $72°$; and finally we completed the pentagon. Euclid first chose the circumcircle and then constructed the pentagon in it.

There were two reasons for the Greek interest in the construction of the regular pentagon. The *pentagram*, or *five-pointed star*, which is obtained by drawing all the diagonals of the regular pentagon, was the mystic symbol of the Pythagorean brotherhood. Second, the construction and computation of the sides of regular polygons inscribed in a fixed circle was a basic step in the construction of tables of the lengths of the chords of a circle (see Section 7-5). These tables were important in Greek astronomy and over the centuries developed into the trigonometric tables that we use today.

In addition to the constructions of the equilateral triangle, the square, and the regular pentagon, Euclid includes those of the regular hexagon, decagon, and quindecagon (15-sided polygon). The latter is constructed by combining an equilateral triangle and a pentagon. We might ask: Why did Euclid not discuss the construction of regular polygons with other numbers of sides? Of course, he could have included constructions for the regular polygons with 8, 12, and 16 sides, but they follow so easily from those of the square and the hexagon that he did not give them. The problem of constructing regular polygons with straightedge and compass only became almost as important as "the three famous problems" of the Greeks. An important contribution to the solution of this problem occurred in 1796 when *Karl Friedrich Gauss* (1777-1855) discovered a construction of the regular 17-sided polygon. Gauss was nineteen years old when he made this discovery, and the excitement that he felt made him decide to become a mathematician. Actually, he became one of the greatest mathematicians of all time.

After discovering the construction of the regular 17-sided polygon, Gauss found a general rule for deciding if, for a given value of n, a regular n-gon can be constructed (with straightedge and compass only). He distinguished two cases: n is a prime number and n is a composite number.

He first proved that if n is a prime number, a regular n-gon is constructible if and only if n can be expressed in the form $2^m + 1$, where m is a natural number. The table shows the values of n for successive replacements of m. Because 3, 5, and 17 are prime numbers, Gauss' theorem tells us that the corresponding regular n-gons are constructible.

m	n	Is n a prime?
1	3	Yes
2	5	Yes
3	9	No
4	17	Yes
5	33	No
.	.	.
.	.	.
.	.	.

Gauss next proved that if n is a composite number, a regular n-gon is constructible if and only if n can be expressed as a product of distinct primes of the form $2^m + 1$ and a power of 2 (which may have the exponent 0). For example, since $15 = 3 \cdot 5$ and 3 and 5 are distinct primes of the form $2^m + 1$, the quindecagon is constructible. However, the regular 9-gon is not constructible because $9 = 3 \cdot 3$, and hence the factor 3 is repeated. The 33-gon is not constructible either, because $33 = 3 \cdot 11$ and 11 is not of the form $2^m + 1$.

Gauss' discovery is remarkable because of the youth of the man who solved a 2000-year-old problem and because number theory was used to solve a geometric problem.

EXERCISES 6-13

1 Proposition II,4 is as follows (in modern notation):
$$(a + b)^2 = a^2 + 2ab + b^2.$$
Euclid's method of proving this proposition is suggested by Figure 6-39.
(a) Give the proof suggested by the figure.
(b) Write a word statement such as Euclid might have used for this proposition.

2 Proposition II,2 is as follows: If a straight line be divided into any two parts the rectangles contained by the whole and each of the two parts are together equal in area to the square on the whole line. Write an algebraic identity that is equivalent to this proposition and draw a geometric diagram that represents it.

3 Show that the length of the side of a regular hexagon inscribed in a circle is equal to the length of the radius of the circle. Construct a regular hexagon. Why are nuts and the heads of bolts either square or hexagonal in shape, not triangular or pentagonal?

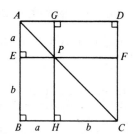

Figure 6-39

4 Construct a 12-sided regular polygon.

5 We showed (in Chapter 4) that an angle of 60° cannot be trisected with straight-edge and compass alone. Use this fact to prove that a regular 9-sided polygon cannot be constructed.

6 (a) Prove that a regular 7-sided polygon cannot be constructed.
   (b) Prove that a regular 14-sided polygon cannot be constructed.

7 There is a legend that the stars in the American flag are five-pointed because Betsy Ross knew an easy process for cutting them out of cloth. Take a sheet of $8\frac{1}{2}$ by 11 paper and fold and cut it as shown in the sequence of diagrams in Figure 6-40. The result is a five-pointed star (or a pentagon or a convex decagon, depending on the angle at which the final cut is made). The method is not exact, however.

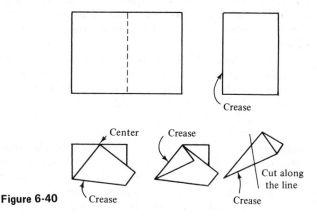

Figure 6-40

8 (a) Carry out the construction of a triangle with angles of 36°, 72°, and 72°, mentioned on page 185. Choose for a a line segment of length 3 inches.
   (b) In a given circle, inscribe a triangle with angles of 36°, 72°, and 72°.
   (c) Construct a regular pentagon in the circle of (b).

9 Look up a construction for a regular 17-sided polygon and carry out the construction.

6-14   NUMBER THEORY IN THE *ELEMENTS*

Euclid is usually thought of as a geometer. However, Books VII, VIII, and IX are devoted to number theory. Some of the important theorems are:

1. there is an infinite number of primes (Proposition IX,20);
2. the Fundamental Theorem of Arithmetic, for a large class of numbers (Proposition IX,14);
3. the Euclidean algorithm for finding the greatest common divisor of two numbers (Proposition VII,2);
4. the theorem that if $2^n - 1$ is a prime, then $2^{n-1} (2^n - 1)$ is a perfect number (Proposition IX, 26; see Section 3-6).

Before we can prove (1), we need Proposition VII,31: *Every composite number is measured by a prime* (by the word "measured" Euclid means "divisible"). Euclid's proof goes as follows: If a is a composite number, then a has a divisor b. The number b is prime or composite. If b is prime, then the theorem is proved. If b is composite, then b has a divisor c which, therefore, is also a divisor of a. The number c is either prime or composite. In the first case, the theorem is proved. In the second case, c has a divisor d which, therefore, is also a divisor of b and hence of a. We can continue in this way. If we did not obtain a prime divisor, we would get an infinite sequence of numbers each of which would be smaller than the previous one. This is not possible with whole numbers. Hence, a also has a prime divisor in the case that b is composite.

Euclid formulates IX,20 as follows: *The primes are more (in number) than any preassigned quantity of primes.* To prove this proposition, Euclid chooses for the "preassigned quantity of primes" the three primes a, b, and c. Now it must be proved that there exists a prime that is different from a, b, and c. Consider the smallest number, s, divisible by a, b, and c (which, of course, is $a \cdot b \cdot c$), and add 1 to it. This gives the number $s + 1$. If $s + 1$ is a prime, then the theorem is proved ($s + 1$ is greater than each of a, b, and c). If $s + 1$ is not a prime, then it is divisible by a prime, d (Proposition VII,31). We must now prove that d is different from a, b, and c. Suppose that $d = a$. Then d does not divide $s + 1$, since a does not divide $s + 1$ (a divides s). Similarly, d is not equal to b or c. Thus, d is a prime different from a, b, and c.

Though the proof was given for the case of three primes, we see immediately that the same reasoning can be used to show that there is always one more prime than any preassigned number of primes. Therefore, it has indeed been proved that the number of primes is infinite.

The modern statement of IX,14, the Fundamental Theorem of

Arithmetic, is: *Every integer greater than 1 is a prime or can be written as a product of primes in exactly one way, except for the order of the factors.* Euclid proves the theorem only for the class of numbers that can be factored into a product of *different* primes. We shall not give a proof of this theorem.

We shall explain Proposition VII,2 as follows. Suppose that we wish to find the greatest common divisor of 72 and 20. Applying the division algorithm to these numbers, we get

$$72 = 20 \cdot 3 + 12. \tag{1}$$

Continuing, we apply the division algorithm to 20 and 12:

$$20 = 12 \cdot 1 + 8; \tag{2}$$

now to 12 and 8:

$$12 = 8 \cdot 1 + 4; \tag{3}$$

then to 8 and 4:

$$8 = 4 \cdot 2 + 0. \tag{4}$$

From equation (1) it follows that the common divisors of 72 and 20 are the same as those of 20 and 12. Similarly, it follows from equations (2), (3), and (4) that the common divisors of 20 and 12, 12 and 8, 8 and 4, and 4 and 0 are the same. Hence, the common divisors of 72 and 20 are the same as those of 4 and 0. The greatest common divisor of 4 and 0 is 4. Hence, the greatest common divisor of 72 and 20 is also 4.

In the equations (1), (2), (3), and (4) the divisors were 20, 12, 8, and 4, respectively. We have seen that the last divisor, 4, is the greatest common divisor of 72 and 20.

This procedure for finding the greatest common divisor of two numbers is called the *Euclidean algorithm*. It can be written in the following form:

$$
\begin{array}{r}
3 \\
20\overline{)72} \\
\underline{60} \quad 1 \\
12\overline{)20} \\
\underline{12} \quad 1 \\
8\overline{)12} \\
\underline{8} \quad 2 \\
4\overline{)8} \\
\underline{8} \\
0
\end{array}
$$

Then 4, the last divisor, is the greatest common divisor of 72 and 20.

The Euclidean algorithm is the basis for the proof of the following modern theorem:

If a and b are numbers with greatest common divisor d, then there are integers x and y such that

$$ax + by = d.$$

By applying this theorem to our example, we find that the equation

$$20x + 72y = 4$$

has a solution in integers. Indeed, by a little experimenting we find that $(-7, 2)$ is a pair of integers that satisfy the equation.

Let us return to the regular quindecagon mentioned in Section 6-13. The quindecagon is constructible because $15 = 3 \cdot 5$ and 3 and 5 are distinct primes of the form $2^m + 1$. We shall use the preceding theorem to show how the construction of the regular quindecagon can be carried out.

The greatest common divisor of 5 and 3 is 1. Hence,

$$3x + 5y = 1$$

has an integral solution. By inspection, we find that $x = 2$, $y = -1$ is a solution. Thus,

$$3(2) + 5(-1) = 1.$$

Dividing by 15, we get

$$\frac{3(2)}{15} + \frac{5(-1)}{15} = \frac{1}{15},$$

or

$$\frac{2}{5} - \frac{1}{3} = \frac{1}{15}. \tag{5}$$

(See Figure 6-41.) Inscribe in a circle an equilateral triangle (with side $\overline{AB}$) and a regular pentagon (with side $\overline{AC}$). Then we have

$$\text{arc } AD = \frac{2}{5} \text{ of the circle}$$

$$\text{arc } AB = \frac{1}{3} \text{ of the circle.}$$

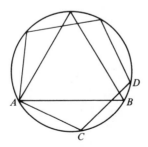

Figure 6-41

Hence, by equation (5),

$$\text{arc } AD - \text{arc } AB = \text{arc } BD$$

$$= \frac{1}{15} \text{ of the circle.}$$

Now we can complete the regular quindecagon.

We have discussed some of the important parts of Euclid's *Elements*. He also wrote other mathematical works. After Euclid, *Archimedes* (287-212 B.C.) and *Apollonius* (225 B.C.) continued the development of mathematics. A number of other mathematicians extended the period of Greek contributions to as late as A.D. 470. Some of the Greek mathematics after Euclid will be mentioned in Chapter 7.

EXERCISES 6-14

1 Show that 496 is a perfect number.

2 Find two integers x and y such that $3x + 17y = 1$. Use this result to tell how to construct a regular 51-gon by combining an equilateral triangle and a regular 17-gon inscribed in the same circle.

3 Use the Euclidean algorithm to find the greatest common divisor of
   (a) 30 and 105   (b) 52 and 2730
   (c) 312 and 396   (d) 24, 180, and 7260

4 A second method for finding the greatest common divisor of two numbers is to write each number as a product of primes and to take the product of the common factors, each to the least of the powers to which it appears. Apply this procedure to (a) and (c) of Exercise 3.

REFERENCES

For a complete translation and analysis of the *Elements*, see reference 21 (Chapter 3).

For a discussion of equivalent forms of the parallel axiom, see

[31] Bunt, Lucas N. H., "Equivalent Forms of the Parallel Axiom," *The Mathematics Teacher*, Vol. 40 (1967), pp. 641-652.

For a discussion of non-Euclidean geometry, see

[32] Eves, Howard, *A Survey of Geometry*, rev. ed. Boston: Allyn and Bacon, Inc. 1972.

[33] Wolfe, H. E., *Introduction to Non-Euclidean Geometry*. New York: Holt, Rinehart and Winston, Inc., 1945.

7

GREEK MATHEMATICS AFTER EUCLID. EUCLIDEAN VS. MODERN METHODS.

7-1   THE SPAN OF GREEK MATHEMATICS

As we move farther from an object, it appears to become smaller. We can no longer see details and we are not certain of all the relations among the parts. This is true in time as well as in space. We refer to Greek mathematics as if it were one whole, existing at one time. Actually, Greek mathematical activity spans over a thousand years, from Thales, about 600 B.C., to Proclus, about A.D. 470.

Archimedes (287-212 B.C.) and *Apollonius* (about 225 B.C.) followed within a century of Euclid. These three were the "all-time greats" of Greek mathematics. Many others approached greatness and are known to us not only for their commentaries, clarifications, and expansions of Euclid's works, but also for individual theorems and extended works of their own. Mathematics students still hear of the sieve of *Eratosthenes* (about 230 B.C.), *Heron's* (about A.D. 75?) formula for the area of a triangle, Diophantine equations (*Diophantus* of Alexandria, about A.D. 250?), the Ptolemaic system of the universe (*Ptolemy*, about A.D. 150) and *Pappus'* (about A.D. 320) theorems on the volume and area of a solid of revolution. Charles Kingsley wrote a novel about *Hypatia*, the first woman mathematician, the daughter of *Theon of Alexandria* (about A.D. 390).

It is as easy to forget the scattered geography of Greek mathematics as it is to forget the long span of years that it covered. Many Greek scholars spent their lives at centers far from the peninsula of modern

Greece. The roll call of mathematicians at Alexandria at the mouth of the Nile included Euclid, Eratosthenes, Heron, Diophantus, Ptolemy, Pappus, and Theon. Pythagoras located at Crotona in Italy, Archimedes on Sicily, and Apollonius at Perga and Pergamum in what is now Turkey.

Mathematical interests and activities changed over these thousand years. Our distance in time from this period is a help here. We can "see the forest," the broad trends, that persons at a particular time might not have perceived. The early Greeks were purists, philosophers, theorists. Later, Archimedes was greatly interested in practical applications of physical principles and made some important inventions, but he believed that his serious contributions were the purely theoretical ones. Heron, still later, wrote about pneumatic and hydraulic machines and about surveying problems and instruments. Ptolemy was as well known for his astronomy, geography, and cartography as for his mathematics.

However, not all the later Greek mathematics was in geometry or directed toward applications. Diophantus' work was of a number theoretic nature, and he even· introduced some rudimentary algebraic symbolism. There was also a revival of interest in the philosophy and number mysticism of the Pythagoreans, called "Neo-Pythagoreanism." It was typified by the work of *Nicomachus of Gerasa* (about A.D. 100) and *Iamblichus* (about A.D. 300).

The scope of this book does not allow the detailed investigation of the extensive mathematical and scientific achievements of the later Greek period. However, we shall discuss some topics that have recently appeared and sometimes still appear in textbooks. After this discussion we shall return to an analysis of the axiomatic method, particularly as it is related to Aristotle and Euclid.

7-2 ARCHIMEDES AND ERATOSTHENES

Archimedes studied in Alexandria but spent most of his life in his homeland, Sicily. He was one of the great mathematicians of all time. Some of his ideas were far ahead of his time. The story of the discovery and deciphering (in 1906) of a letter from Archimedes to his old friend Eratosthenes at Alexandria is like a detective story. In the letter he told Eratosthenes of the method that he used to *discover* many of his theorems about areas, volumes, and centers of gravity. Archimedes' method of discovery was close to the methods used in the seventeenth century by Newton and Leibniz in developing the calculus. However, when he published his results he used classical Greek proofs and did not explain how he had *discovered* his theorems. His results and proofs were preserved and used by mathematicians and astronomers for centuries. Exercise 9 of Exercises 4-7 gives an example of his effect on Johannes Kepler. If his method for discovery had been known and understood during his lifetime, mathematics might be much further advanced today!

Archimedes derived a formula for the volume of a sphere by showing that this volume is equal to that of the cylinder circumscribed around the sphere minus the volume of the cone inscribed in the cylinder (see Exercise 2 of Exercises 7-2). He is supposed to have asked that his diagram for this theorem be carved on his tombstone. The method of proof involved comparing slices of the three solids. This method was much like that of *Cavalieri* (1598-1647).

There are many legends about Archimedes as a person and about his mechanical inventions. Some of the war machines that he invented were used to defend his home town, Syracuse, against the Roman invaders from the sea. These machines included levers, pulleys, grappling devices, and burning mirrors. His spiral curve could be used to trisect angles (see Exercise 3 of Exercises 7-2) and today gives a shape for the casing of a centrifugal pump. His *Sand-Reckoner* was really a treatise on how one could write and compute with large numbers, such as would give the number of grains of sand in the universe. By inscribing and circumscribing a 96-sided regular polygon in and about a circle he computed that π is between $3\frac{1}{7}$ and $3\frac{10}{71}$. We have already discussed a device, different from the spiral curve just mentioned, that he used for trisecting an angle (see Section 4-6). His name is given to an axiom that is of fundamental importance in modern developments of the real-number system and the topology of the real line.

Eratosthenes is noted for his determination of the circumference of the earth. The *sieve of Eratosthenes* is a simple method for finding prime numbers. Eratosthenes considered the sequence of integers 2, 3, 4, 5, ... Following 2, the first prime, he crossed off every other integer, that is, 4, 6, 8, 10, ... Following 3, the second prime, he crossed off every third number, 6, 9, 12, 15, ... After 5 he crossed off every fifth number, after seven every seventh, and so on. The remaining numbers,

$$2, 3, 5, 7, 11, 13, 17, 19, 23, \dots ,$$

are prime.

Mathematicians have found other ways of looking for primes, but no practicable procedure is known for determining if a given large number is prime nor has a formula been found for producing primes. Euclid showed that there is no largest prime. Gauss and others have shown that the primes are less dense in the sequence of integers as the integers get larger.

EXERCISES 7-2

1 (a) Use the sieve of Eratosthenes to find the primes up to 100.
  (b) Explain why it is only necessary to cross off through the multiples of 7 to find the primes less than 100.

(c) To find the primes less than n, what is the largest number the multiples of which must be crossed off?

2 Figure 7-1 shows Archimedes' cylinder, sphere, and cone, cut by a plane P.

(a) Show that for the circles cut off from P: area cut off by sphere = area cut off by cylinder — area cut off by cone.

(b) Use the result in (a) to prove: volume sphere = volume cylinder — volume cone. (Hint: Assume that the cylinder is a stack of thin disks.)

(c) Check the result in (b) by using modern formulas for the volume of the sphere, the cylinder, and the cone.

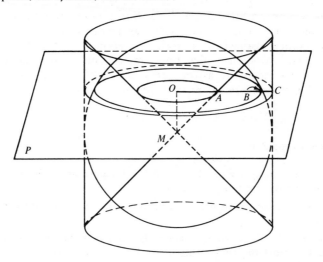

Figure 7-1

3 The spiral of Archimedes is defined in polar coordinates by the equation $\rho = a\theta$.

(a) Choose a unit of length and sketch the spiral of Archimedes.

(b) Show how the spiral can be used to trisect a given angle.

4 Read and report on a biography of Archimedes.

5 Read and report on a biography of Eratosthenes. Explain, with a diagram, his method for determining the radius and the circumference of the earth.

7-3 APOLLONIUS OF PERGA

Apollonius' most important contribution was his work on the conic sections. We have seen earlier (see Section 4-5) that the problem of duplicating the cube was reduced, by Menaechmus, to finding the intersection of two parabolas. The trisection problem can be solved by using a hyperbola (see Exercise 15 of Exercises 4-7). Euclid wrote a work on conic sections which we no longer have. However, Apollonius organized, synthesized, and greatly extended earlier work. Of his seven books on the conics, the first four are assumed to be based on Euclid's work. Apollonius originated the names *ellipse, hyperbola*, and *parabola* for the conic sections and developed constructions for tangents and normals to them.

Apollonius solved a set of 10 tangency problems which have intrigued geometers ever since. The problems deal with constructing a circle tangent to three given circles, where the given circles are permitted to degenerate independently into points (by shrinking) or lines (by expanding infinitely). For example: given three points, find the circle through them; or, given three lines, find the circles tangent to all three; or, given a point, a line, and a circle, find a circle passing through the point, tangent to the line, and tangent to the circle.

Apollonius also wrote a work on verging constructions (see Section 4-6). His work on conics, however, was his greatest achievement. It was known to such astronomers and mathematicians as Kepler, Galileo, Copernicus, Descartes, and Newton.

EXERCISES 7-3

1 Define the 10 tangency problems by listing the possible combinations of three elements drawn from points, lines, circles. Sketch a figure for each.

2 Given: three noncollinear points A, B, and C. Construct a circle through A, B, and C.

3 Given: three lines, no two of them parallel and not all three passing through one point. Construct the circles tangent to the three lines. (Hint: There are four solutions.)

4 Given: three lines, two of which are parallel. Construct the circles tangent to the three lines.

5 Under what conditions is there no solution to the problem of finding a circle tangent to three given lines? Are there three lines such that one and only one circle can be drawn tangent to them?

6 One of the tangency problems is: Given three circles, construct a circle tangent to them. There are many possible arrangements of the given circles.
   (a) Sketch an arrangement where no solution exists.
   (b) Sketch an arrangement where eight solutions exist.

7 The following theorem has been attributed to Apollonius (see Figure 7-2). If $\overline{CM}$ is a median of $\triangle ABC$ and if $AB = c$, $BC = a$, $CA = b$, and $CM = m$, then

$$a^2 + b^2 = 2m^2 + 2\left(\frac{c}{2}\right)^2.$$

   (a) Prove this formula. (Hint: Draw the altitude from C and apply the Pythagorean theorem.)
   (b) Show that the theorem reduces to the Pythagorean theorem if $\angle ACB$ is a right angle.

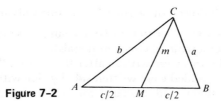

Figure 7-2

7-4 HERON OF ALEXANDRIA AND DIOPHANTUS

The name of Heron is associated with a formula for the area of a triangle in terms of the lengths of its sides:

$$A = \sqrt{s(s-a)(s-b)(s-c)},$$

where $s = \frac{1}{2}(a + b + c)$.

Although this formula is probably actually due to Archimedes, it is natural that it should be associated with Heron, since his proof is the oldest proof that we have. Heron was interested in surveying, mensuration, and optics. He wrote about water- and steam-powered apparatus and invented the first jet engine. His interest in such topics represents a change from the philosophical interests of the earlier Greek mathematicians.

There is some doubt about the time that Diophantus lived, but most scholars place him in Alexandria about 250 A.D. The *Arithmetica* of Diophantus is a collection of about 150 solved problems. Each solution depended on a polynomial equation, sometimes in more than one unknown, for which a rational solution was to be found. This restriction to rational solutions is the origin of the modern custom of giving the name "Diophantine equations" to equations in several variables for which integral solutions are required.

Perhaps Diophantus' most significant achievement was his contribution to algebraic symbolism. He used the following abbreviating symbols for powers of the unknown (x):

$$\overset{\circ}{M} \qquad \text{unity } (x^0)$$

$$\varsigma \qquad \text{number } (x)$$

$$\Delta^Y \qquad \text{square } (x^2)$$

$$K^Y \qquad \text{cube } (x^3)$$

$$\Delta^Y\Delta \qquad \text{square-square } (x^4)$$

$$\Delta K^Y \qquad \text{square-cube } (x^5)$$

$$K^Y K \qquad \text{cube-cube } (x^6)$$

Diophantus followed these rules for writing polynomials:

1. The coefficients were written in Greek numerals after the unknown (see page 67 for a table of Greek numerals).
2. All subtracted terms were written after the symbol "Λ."
3. The terms to be added were written side by side without an addition sign between them, and similarly for the terms to be subtracted.

EXAMPLES

$$\Delta^Y \bar{\gamma} \overset{\circ}{M} \overline{\iota\epsilon} = 3x^2 + 15$$

$$\Delta^Y\Delta \ \bar{\alpha} \overset{\circ}{M} \ \mathcal{M} \ \Lambda\Delta^Y\overline{\xi} = x^4 - 60x^2 + 900$$

$$K^Y \bar{\alpha} \ \varsigma \ \bar{\eta} \ \Lambda \ \Delta^Y\overline{\epsilon} \overset{\circ}{M} \bar{\beta} = x^3 - 5x^2 + 8x - 2.$$

The examples show that Diophantus could write polynomials very nearly as concisely as we can.

Because of the originality of his work and the availability of the *Arithmetica* in Western Europe in the time of the Renaissance and later, Diophantus has had a greater influence on algebraic notation and on the theory of numbers than any of the other Greek mathematicians.

EXERCISES 7-4

1 Heron's formula can be regarded as a special case of the following formula for the area of a convex quadrilateral inscribed in a circle due to *Brahmagupta* (India, seventh century A.D.):

$$A = \sqrt{(s-a)\ (s-b)\ (s-c)\ (s-d)},$$

where A is the area; $a, b, c,$ and d are the lengths of the sides; and $s = \frac{1}{2}(a+b+c+d)$. Show that if points A, B, and C in Figure 7-3 remain fixed while D approaches A along the circle, then the quadrilateral approaches a triangle and Brahmagupta's formula approaches Heron's formula.

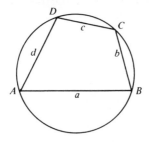

Figure 7-3

2 Prove Heron's formula.

3 The following problem is from the *Arithmetica* of Diophantus: Find two numbers such that their sum is 20 and the sum of their squares is 208. In solving the

problem, he assumed the numbers to be $10 + x$ and $10 - x$ (since the sum of these two numbers is 20; see Example 2 on page 52 for a similar method used by the Babylonians). Complete the solution.

4 Another problem from the *Arithmetica* is: Find two numbers such that their sum is 10 and the sum of their cubes 370. Solve this problem by using the procedure of Exercise 3.

5 Look up a discussion of the symbolism of Diophantus in reference 36 (this chapter) or reference 7 (Chapter 1).

6 Translate into modern notation:

(a) $\Delta^Y \, \overline{\xi} \, \mathring{M} \, \overline{,\beta \, \phi \, \kappa}$

(b) $\Delta^Y \Delta \, \overline{\alpha} \, \mathring{M} \, \overline{\lambda} \, \varsigma \, \wedge \, \Delta^Y \, \overline{\iota \, \beta}$

(c) $\Delta K^Y \overline{\gamma} \, \Delta^Y \overline{\iota \, \gamma} \, \varsigma \, \overline{\mu \, \epsilon} \, \wedge \, \Delta^Y \Delta \, \overline{\delta} \, \mathring{M} \, \overline{\sigma \, \kappa \, \eta}$

7 Translate into the notation of Diophantus:
(a) $32x + 9$
(b) $41x^2 - 3$
(c) $2x^3 - 3x^2 + 48$

7-5 PTOLEMY AND PAPPUS

There were many other Greek mathematicians. In later times they became more practical and computational in their work, but they continued to anticipate later work in the calculus and in analytic and projective geometry, while extending the ideas of their forerunners.

We are little informed about the life of Ptolemy but we know that he lived in Alexandria, where he made astronomical observations, in the second century A.D. His major work, the *Almagest* (Arabic: the greatest), has been preserved. Book I of the *Almagest* contains the theorems necessary to build up a table of the lengths of the chords of a circle in terms of the angle subtended by the chord. This table is equivalent to a modern table of sines. Ptolemy worked out the table for each angle from $\frac{1}{4}°$ to $90°$ in steps of $\frac{1}{4}°$. This table was one of the mathematical tools that he used to develop his system of the universe.

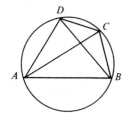

Figure 7-4

Ptolemy's theorem, which was fundamental to the construction of the table, is as follows (see Figure 7-4): *If ABCD is a (convex) quadrilateral inscribed in a circle, then*

$$AB \cdot CD + AD \cdot BC = AC \cdot BD.$$

The reader will be asked to prove this theorem in Exercise 6 of Exercises 7-5.

Ptolemy used the above theorem to establish the equivalent of several modern trigonometric formulas. For instance, he was able to prove the equivalent of the following formula:

$$\sin(\alpha - \beta) = \sin \alpha \cos \beta - \cos \alpha \sin \beta$$

(see Exercise 8 of Exercises 7-5). He used these formulas to build up his table.

The last of the great mathematicians at Alexandria was Pappus. Much of Pappus' principal work, called the *Collection*, has been preserved and it is from this source that we know of some of the contributions of Archimedes, Euclid, Apollonius, and others. Pappus' own contributions include discussion, at a high level, of the trisection problem, the rudiments of analytic geometry, the isoperimetric problem (in which he analyzed the efficiency of the hexagonal cell of the honeybee), and the conic sections (including the focus-directrix property).

The theorem of which Pappus was most proud is the following: If a closed plane curve is revolved around a line in its plane not passing through the curve, then the volume of the solid generated is found by taking the product of the area bounded by the curve and the distance traveled by the centroid of the area (see Figure 7-5). He also gave the

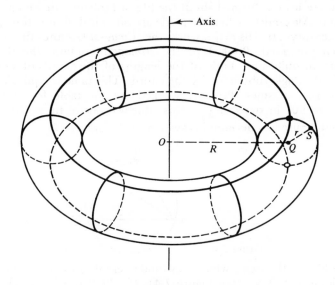

Figure 7-5

companion theorem: If a curve is revolved around a line in its plane not passing through the curve, then the area of the surface generated is the product of the length of the curve and the distance traveled by the centroid of the curve. The reader will be asked to apply these theorems in Exercise 2 of Exercises 7-5.

We have repeatedly pointed out contrasts and connections between Greek geometry and modern views of geometry. We shall examine these questions a little more deeply in the next section before we take leave of Greek geometry.

EXERCISES 7-5

1 Nichomachus extended a theorem of the Pythagoreans on square numbers to cubic numbers. The theorem is: The sum of the first n odd numbers is equal to n^2 (see Exercise 7 of Exercises 3-6). Nicomachus saw that one can obtain the successive whole number cubes (1^3, 2^3, 3^3, ...) by adding successive sequences of odd numbers (the first odd number, the next two odd numbers, the following three odd numbers, and so on) thus:

$$1 = 1^3$$
$$3 + 5 = 8 = 2^3$$
$$7 + 9 + 11 = 27 = 3^3$$
$$\text{etc.}$$

Check Nichomachus' theorem for $n = 5$ and $n = 7$.

2 Figure 7-5 represents a *torus*. This is a doughnut-shaped solid obtained by rotating a circular region of radius r about a line at distance R from its center.
   (a) Use Pappus' theorem to find the volume of the torus.
   (b) Find the volume of a torus that has a hole of diameter $\frac{1}{2}$ inch and a maximum distance across the torus of $2\frac{1}{2}$ inches.
   (c) Find the surface area of a torus in terms of r and R.
   (d) Find the surface area of the torus in (b).

3 The following theorem is due to Pappus: If A, C, and E are three points on one line, and B, D, and F three points on a second line, then the points of intersection of $\overleftrightarrow{AB}$ and $\overleftrightarrow{DE}$, $\overleftrightarrow{BC}$ and $\overleftrightarrow{EF}$, and $\overleftrightarrow{FA}$ and $\overleftrightarrow{CD}$ are collinear. Verify this theorem by construction.

4 Show that if an inscribed quadrilateral is a rectangle, then the Pythagorean theorem appears as a special case of Ptolemy's theorem.

5 Pappus proved the following theorem (see Figure 7-6): If $\overline{AB}$ is the diameter of a circle with center O, C a point of $\overline{AB}$ different from O, $\overline{DC} \perp \overline{AB}$, and $\overline{CE} \perp \overline{OD}$, then OD, CD, and DE are the arithmetic mean, the geometric mean, and the harmonic mean, respectively, of AC and CB.
   (a) Prove the theorem (see Section 3-5 for a definition of the means).
   (b) Prove: $DE < CD < OD$.

6 See Figure 7-4. Prove Ptolemy's theorem. (Hint: Find E on $\overline{AC}$ such that $\angle ABE \cong \angle DBC$ and consider the similarity of $\triangle ABE$ and $\triangle DBC$ and of $\triangle EBC$ and $\triangle ABD$.)

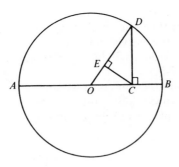

Figure 7–6

7   Prove: If in a circle with radius of length R a chord of length k subtends an inscribed angle of measure α, then $k = 2R \sin \alpha$.

8   Let quadrilateral $ABCD$ be inscribed in a semicircle with radius R such that $\overline{AB}$ is the diameter. Let $m\angle BAD = \alpha$ and $m\angle BAC = \beta$. Prove by applying Ptolemy's theorem that $\sin(\alpha - \beta) = \sin\alpha \cos\beta - \cos\alpha \sin\beta$. (Hint: See Exercise 7.)

7-6   REVIEW OF THE GREEK METHOD

According to Aristotle, the deductive method differs from others by the following characteristics:

1. The concepts are defined. They are not merely described more or less vaguely with their meaning made clear by examples from experience.
2. The propositions are proved. They are not merely made plausible or thought to be correct on account of empirical data.

Without a starting point, however, one cannot give definitions or proofs. Hence, Aristotle starts from

1. Some statements expressing the existence of certain undefined fundamental concepts (special notions),
2. Some statements by which the properties of these fundamental concepts are laid down (also special notions), and
3. Some universally true statements that may be used for the deduction of new propositions (common notions).

When we discussed the *Elements* of Euclid, we found a clear similarity between the views of Aristotle and the method of Euclid.

1. Euclid starts with some fundamental concepts: point, straight line, and circle. In Postulates I through III, the existence of these concepts is laid down.

2. In Definitions I through IV and XV, we find a listing of some of the fundamental properties of these concepts. Their aim is to elucidate the meaning of these concepts.

3. The Euclidean axioms state the general properties that are used in the proofs.

So far the similarity is clear. However, the Euclidean system is also based on:

4. Postulates IV and V. These postulates do not serve to lay down the *existence* of the fundamental concepts or to elucidate their *meaning*. Neither do they have a general character. Hence, they belong to none of the three means accepted by Aristotle for laying down the foundations of a deductive system. Euclid accepted them as correct, because they appeared to be essential for the construction of his system, but he did not succeed in giving a proof of them.

7-7 OBJECTIONS TO THE EUCLIDEAN SYSTEM

In several places in Chapter 6 we mentioned objections to the Euclidean system that can be made from a modern point of view. We state them once more.

1. The construction of the system starts from fundamental knowledge that is not clear cut. To understand the axioms, one has to know the meaning of "equal things," "adding," "subtracting," "remainder," "whole," and "part."

2. The Definitions I through IV and XV suffer from the same defect. The meaning of several terms occurring in these definitions is not given: for example, the meaning of "part" (Definition I), "breadthless," "length" (Definition II), "end" (Definition III), "to lie evenly with" (Definition IV), "plane figure," "enclose," "are equal to" (Definition XV). Hence, the meaning of the concepts point, straight line, and circle is not properly explained by these definitions.

It turns out that Euclid does not use these definitions anywhere in constructing his system. Hence they could have been left out. (The same is true for the Definitions V through VII.) Moreover, in Definition VIII the plane angle is defined in an inadmissable way by using the term "inclination."

3. In some proofs, certain properties of figures are taken for granted; that is, these properties are not proved. We recall the following instances:

a. The existence of a point of intersection of the two circles in the proof of Proposition 1.

b. The existence of a point of intersection of a circle and a line through the center in the proof of Proposition 2.

c. The use of motions in the proof of the Propositions 4 and 8 (SAS and SSS), although the meaning of "motion" has not been defined.

d. The existence of a point of intersection of the lines $\overleftrightarrow{GK}$ and $\overleftrightarrow{EF}$ in the proof of Proposition 30.

Hence, Euclid accepts many more unproved properties than those stated in Postulates IV and V.

If we want to improve Euclid's system, it is clear that we have to make two demands:

1. A sharp formulation of the starting point.
2. The avoidance of any gap in a proof; that means the avoidance of any use of properties the correctness of which is derived from a figure only.

7-8   THE MEANING OF DEDUCTION

In order to show how Euclid's system could be improved, we shall first explain the meaning of deduction (proof).

EXAMPLE 1

We start from the following two *statements*:

> Socrates is a man
> all men are mortal.

From this we can draw a *conclusion*:

> Socrates is mortal.

With reference to this example we ask a question that looks very strange at first sight but which is of fundamental importance. The question is: Is it necessary to know the meaning of "man," "mortal," and "Socrates" to draw the conclusion?

Suppose that a visitor has recently arrived in our country. His knowledge of our language is very poor but he knows a few words. The word "mortal" is unknown to him. We tell him:

> Socrates is a man
> all men are . . . . (an incomprehensible word)

and we ask him to draw a conclusion. He then has sufficient data to answer:

> Socrates is . . . . (the same word).

If he does not know the word "man," he can still understand that

> Socrates is a — — — —
> all — — — — are . . . .

hence,

> Socrates is . . . .

is a *valid argument*.

And even if the word "Socrates" is incomprehensible to him, he will still be able to understand that

$$-.-.- \quad \text{is a} \quad - - - -$$
$$\text{all} \quad - - - - \quad \text{are} \quad \ldots .$$

hence,

$$-.-.- \quad \text{is} \quad \ldots .$$

is a valid argument.

Thus, he can judge the validity of this argument without knowing the meaning of the words that indicate the things and without understanding the properties to which the statements refer.

Remarks

1. In referring to statements we use the adjectives "true" and "false." In referring to arguments we use the adjectives "valid" and "invalid."

2. In order to verify the validity of an argument, it is, of course, necessary that the meaning of at least some terms is known, in this case the meaning of "all," "are," and "is a." The science that concerns (among other things) those terms the meaning of which must be known in order to judge the validity of an argument is called *logic*. In this science the meaning of such terms as "is a," "all," "exists," "some," "consequently," "and," "or," and "not" is studied.

EXAMPLE 2

The electric light at home suddenly fails. I want to know if the cause is a short circuit. I know:

if there is a short circuit, then a fuse is blown.

I check the fuses and find:

no fuse is blown.

Hence, I conclude:

there is no short circuit.

In this example we deal with the statements

there is a short circuit   (statement A)
a fuse is blown   (statement B).

We know:

if A, then B

and we know   B is not true.

Hence, we conclude:

A is not true.

To draw this conclusion it is superfluous to understand the contents of the statements A and B. It is not necessary to know what is meant by the words "short circuit," "fuse," and "is blown."

These two examples illustrate what is meant by drawing a conclusion. *To draw a conclusion means to deduce a statement from other statements in such a way that it is not necessary to know the meaning of the concepts found in these statements. This deduction must be carried out in such a way that, if the original statements are true, the deduced statement is also true.*
The original statements are called *premises*; the deduced statement is called the *conclusion*. For example,

all A are B
all B are C   (premises)

hence,

all A are C   (conclusion)

is a valid argument. Whatever the meaning of A, B, and C, the conclusion is a true statement if the premises are true statements. (Replace, for instance, A by "cows," B by "mammals," and C by "living beings.")
If we start with false premises, it may easily occur that we deduce a false statement. For instance,

all male sparrows are sparrows
all sparrows are mammals   (premises)

hence,

all male sparrows are mammals   (conclusion).

We applied the above scheme; hence, the conclusion is drawn in the correct way. (We were not allowed to reflect on the meaning of the words "male sparrow," "sparrow," and "mammal.") The second premise, however, is false, so that we need not be astonished that the conclusion is false, too. From false premises we deduced in a correct way a false conclusion.

An example of a wrong argument is

$$\left.\begin{array}{l} \text{all sparrows are birds} \\ \text{all male sparrows are birds} \end{array}\right\} \quad \text{(premises)}$$

hence,

some sparrows are male sparrows   (conclusion).

The conclusion looks trustworthy; it is a true statement. This statement, however, has been deduced by an invalid argument. Consider the scheme

$$\left.\begin{array}{l} \text{all A are C} \\ \text{all B are C} \end{array}\right\} \quad \text{(premises)}$$

hence,

some A are B   (conclusion).

We can easily show that even if the premises are true, the conclusion need not be true. Choose for instance for A: cows, for B: horses, and for C: mammals; we then get

$$\left.\begin{array}{l} \text{all cows are mammals} \\ \text{all horses are mammals} \end{array}\right\} \quad \text{(premises)}$$

hence,

some cows are horses   (conclusion).

And this is definitely not true.

The last two examples are special cases of the above scheme, in which the statements have the structures

all A are C
all B are C
some A are B.

In the first example the conclusion is true; in the second example the conclusion is false. Since arguments of this form can lead to a false conclusion when we start from true premises, all arguments of this form are called *invalid*.

We see that, when judging the validity of an argument, we only need to pay attention to the *structure of the statements* in the argument. We can ignore the *meaning of the concepts* to be found in the statements.

A science in which the truth of every new statement is deduced exclusively by drawing conclusions from statements that already have been accepted as true is called a *deductive science*.

The empirical sciences (for example, physics) are essentially different from the deductive sciences. In physics we start from observations. It is true that a theory is built up later, but the results of this theory must be in accordance with the observations; if not, the theory is rejected.

7-9   EUCLID'S SYSTEM IS NOT PURELY DEDUCTIVE

We shall analyze once more the proof of Proposition 1 (see Section 6-5). This proof is based on the following two postulates:

> If P and Q are points, then there exists a line segment $\overline{PQ}$.   (1)

> If P and Q are points, then there exists a circle with center P and radius $\overline{PQ}$.   (2)

Euclid reasons as follows.

1. A and B are points;
   hence, there exists a circle with center A and radius $\overline{AB}$ (according to postulate (2)).
2. B and A are points;
   hence, there exists a circle with center B and radius $\overline{BA}$ (according to postulate (2)).
3. These circles have a point of intersection C (this appears from the figure).
4. A and C are points;
   hence, there exists a line segment $\overline{AC}$ (according to postulate (1)).
5. B and C are points;
   hence, there exists a line segment $\overline{BC}$ (according to postulate (1)).

Steps 1, 2, 4, and 5 clearly contain conclusions. Even if we did not know the meaning of "point," "line segment," "circle," "center," and "radius," we would be able to see the truth of the obtained results. Step 3, however, is not a logical conclusion. Only by referring to the figure can we believe the statement. If we did not know what a circle is, we would not be in a position to conclude from the data that there exists a point of intersection C.

Consequently, we see that the discovery of a gap in the reasoning means that we have proved that Euclid's system is not purely deductive. *For a purely deductive development of geometry, it is necessary to mention explicitly in the postulates everything that cannot be proved without the aid of a figure.* This is exactly what Euclid did for some properties in Postulates IV and V.

Now it also becomes clear why Definitions I through IV do not play a role in Euclid's system. For we realize that in our deductions we may not use the meaning of the words "point" and "line segment." Hence, it was superfluous to lay down the meaning of these words, which was the purpose of Definitions I through IV.

7-10   HOW IS GEOMETRY BUILT UP PURELY DEDUCTIVELY?

The first purely deductive construction of geometry was not achieved until more than 2000 years after Euclid. The designer of this system was *David Hilbert*, a German mathematician (1862-1944). His system dates from 1899 and was gradually improved in successive editions of his book (English title: *Foundations of Geometry*; see reference 39 of this chapter).

Hilbert started from a number of basic properties, which he accepted without proof. He called them *axioms*. (Here, "axiom" does not have the same meaning as in Euclid's system. Hilbert's axioms are more comparable to Euclid's postulates.) All further geometric propositions were derived from these axioms in a purely deductive way.

It would be going too far here to study Hilbert's geometrical system. However, we can get insight into a purely deductive construction of geometry in a much simpler way. For that purpose we shall start from a simple set of axioms and be satisfied by deducing some propositions.

We choose as axioms:

AXIOM 1   *Through every two points passes a line.*
AXIOM 2   *Through two points passes not more than one line.*
AXIOM 3   *There exist three points such that a line passing through two of them does not pass through the third.*
AXIOM 4   *If the line l does not pass through the point P, then there exists a line parallel to l which passes through P.*
AXIOM 5   *If the line l does not pass through the point P, then there exists not more than one line parallel to l and passing through P.*

From these axioms we now want to deduce properties. We know that deduction takes place without consulting the figure, hence without using any knowledge of the geometric meaning of the words occurring in the axioms. First let us list the terms the meaning of which we may not use when drawing conclusions. These are:

point,
line,
passes through.

If we use the concepts point and line, we think of geometric figures. But we may not use this geometric meaning of "point" and "line" when drawing conclusions. And if we use the relation "passes through" with reference to a line and a point, we visualize a line passing through that point. We may not use this intuitive meaning either when drawing conclusions.

We call "point" and "line" the *basic concepts* of the system, "passes through" the *basic relation*. We must give a definition of all the remaining concepts and relations occurring in the system. In giving such definitions today, we do not stick completely to the Aristotelian requirements. For instance, we do not require that the existence of every defined notion be proved. But we do require that the meaning of the new concepts and the new relations be explained by using nothing else but basic concepts, basic relations, and previously defined concepts and relations. In this way, for example, we can define the relation "is parallel to" as follows:

DEFINITION   *Line l* is parallel to *line m means that there is no point through which l and m both pass.*

We also need another relation, which we shall define, the relation "intersects."

DEFINITION   *Line l* intersects *line m means that there is a point through which l and m both pass. This point is called the* point of intersection *of l and m.*

According to the Axioms 1 and 2, one and no more than one line passes through A and B.

DEFINITION   *The line through the points A and B is called the line* AB.

From our set of axioms we shall draw a number of conclusions. In doing so, we shall aim at a modest goal, the proof of the proposition: *Given a line, there exist at least three points through which the line passes.*

PROPOSITION 1   *There exist at least three lines.*

Proof   According to Axiom 3, there exist three points such that a line passing through two of them does not pass through the third. Suppose that A, B, and C are three such points. Then lines BC, CA, and AB are three different lines, for if BC were the same as CA, for example, then BC would pass through A, B, and C. This is contrary to the assumption that no line passes through A, B, and C.

Remark In a proof we are not allowed to use the graphical meaning of "point," "line," and "passes through"; that is, we may not use the figure. Yet it is often easier to follow the reasoning if a figure is drawn. Therefore, Figure 7-7 ought to be considered only as a help in following the proof. We have not referred to this figure; it is essentially superfluous. A similar remark applies to each of the other figures of this section.

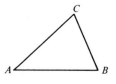

Figure 7-7

PROPOSITION 2 *There exist at least six lines.*

Proof Let A, B, and C be three points such that there is not one line passing through all three. Draw the lines $l_A \parallel BC$, $l_B \parallel CA$, and $l_C \parallel AB$ through A, B, and C, respectively (Axiom 4). We shall now prove that the lines AB, BC, CA, l_A, l_B, and l_C are different lines.

1. We know already that the lines AB, BC, and CA are different lines (see the proof of Proposition 1).
2. l_A and BC are different lines, for l_A passes through the point A and BC does not. Similarly, we prove that l_B and CA, and l_C and AB are different lines.
3. l_A and AB are different lines, for l_A is parallel to BC and AB intersects BC. Similarly, we prove that l_A and CA, l_B

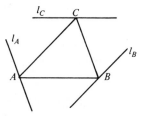

Figure 7-8

and AB, l_B and BC, l_C and CA, and l_C and BC are different lines.
4. l_A and l_B are different lines, for l_A is parallel to BC, and l_B intersects BC. Similarly, we prove that l_B and l_C, and l_C and l_A are different lines.

This proves that the six lines mentioned above are six different lines.

Remark Henceforth, in each proof the letters have the same meaning as in the proof of Proposition 2. That is, A, B, and C will always be three points such that there is not one line passing through all three, l_A will be a line through A parallel to BC, l_B a line through B parallel to CA, and l_C a line through C parallel to AB.

PROPOSITION 3 *There exist at least four points.*

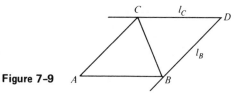

Figure 7-9

Proof   First we prove that l_B intersects l_C.

AB passes through A

l_B does not pass through A $\Big\}$ $\therefore l_B$ is different from AB.

l_B and AB both pass through B

$AB \parallel l_C$ $\Big\}$ $\therefore l_B$ is not parallel to l_C
 (Axiom 5).

Hence, l_B and l_C are intersecting lines. Let D be the point of intersection.

Next we prove that D is distinct from A, B, and C.

l_C passes through D

$l_C \parallel AB$ $\Big\}$ $\therefore D$ is distinct from A and B.

In the same way we prove that D is distinct from $(A$ and$)$ C. Hence, D is distinct from A, B, and C.

Remark   We proved that l_B and l_C intersect. In the same way we can prove:

l_C and l_A intersect in a point E.
l_A and l_B intersect in a point F.

At this point it seems as though we have proved that there exist six different points. This is not true, however, as long as we are not sure that D, E, and F are different points. If we look at Figure 7-10 this seems to be evident. But such an argument has no value since we agreed that only a deductive proof should count. The remarkable thing is that it is not possible to derive from our five axioms the fact that D, E, and F are different points.

To obtain our aim stated on page 212 (namely, to prove the theorem: Given a line, there exist at least three points through which the line passes), it is necessary to add a new axiom, according to which the points D, E, and F are different.

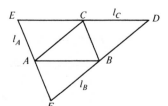

Figure 7-10

AXIOM 6   *If there exists no line passing through A, B, and C, and if*

> *a line $l_A \parallel BC$ passes through A*
> *a line $l_B \parallel CA$ passes through B*
> *a line $l_C \parallel AB$ passes through C*
> l_B *and* l_C *intersect in D*
> l_C *and* l_A *intersect in E*
> l_A *and* l_B *intersect in F*

then D, E, and F are three different points.

COROLLARY   *There exist at least six points.*

PROPOSITION 4   *There exist at least four lines passing through A.*

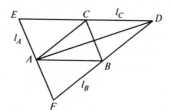

Figure 7-11

Proof   The line AD passes through A and D. Hence, the lines AB, AC, AD, and AE pass through A. We shall prove that these lines are distinct lines.

We only need to show that AD is different from each of the lines AB, AC, and AE, because we already proved in Proposition 2 that AB, AC, and AE (that is, l_A) are different lines.

1. We prove: AD is different from AB.

> $AB \parallel ED$ (l_C) $\left.\vphantom{\begin{array}{c}a\\b\end{array}}\right\}$ $\therefore AB$ does not pass through D.
> ED passes through D

> AB does not pass through D $\left.\vphantom{\begin{array}{c}a\\b\end{array}}\right\}$ $\therefore AB$ and AD are different lines.
> AD passes through D

2. AD is different from AC. This is proved similarly.
3. We prove: AD is different from AE.

> ED $(l_C) \parallel AB$ $\left.\vphantom{\begin{array}{c}a\\b\end{array}}\right\}$ $\therefore ED$ does not pass through A.
> AB passes through A

Hence, no line passes through E, D, and A, and hence AD and AE are different lines.

PROPOSITION 5   *Given a line, there exist at least three points through which the line passes.*

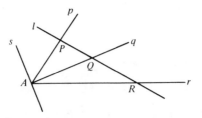

Figure 7-12

Proof   Let A, B, and C be three points satisfying the condition that there is no line passing through all three of them, and let l be an arbitrary line. Since, by assumption, l cannot pass through all three of the points A, B, and C, there is at least one of these points through which l will not pass; let this point be A.

Proposition 4 says that there are four lines passing through A. We call these lines p, q, r, and s, respectively. According to Axiom 5, l is parallel to at most one of these lines; hence, l has a point of intersection with at least three of them. We shall assume that l intersects p, q, and r in the points P, Q, and R, respectively.

Line l does not pass through A, hence A is different from P, Q, and R.

Further, P is different from Q, for if P were the same point as Q, then AP would be the same line as AQ (Axiom 2), that is, the same line as q, which is not true.

In the same way we prove that P and R, and also Q and R, are different points.

Hence, there are three different points through which l passes.

Continuing in this way, it would be possible to build up plane geometry, but then we would need more axioms. However, we only intended to give an insight into the axiomatic-deductive method. The preceding theory may have served that purpose.

Arguments such as those in the preceding paragraphs require a degree of abstraction that was not attained by the Greeks. Nevertheless, if we compare the Greek method with the modern method, we are impressed with the similarities between them. The Greeks understood that only purely deductive proofs were allowed. In applying this principle, they sometimes failed, as is evident from the flaws in their proofs of theorems, which we have pointed out several times. Furthermore, they did not succeed in finding a starting point that would be satisfactory from a modern point of view. The main reason for this is that they did not see that the meaning of the basic concepts is of no importance for the construction of the system, and hence that there is no need for stating the meaning of these concepts explicitly.

It would be unfair to reproach them for this. It is rather more becoming to express our great appreciation for their very important contribution to the development of geometry as a deductive science.

7-11 A FOUR-POINT SYSTEM

We return to the remark that we made in relation to Figure 7-10. We not only mentioned the fact that we had not proved that the points D, E, and F are different, but, more strongly, we said that it is *impossible* to prove this from the first five axioms.

This may sound strange: we do not say that some particular assertion is not true, but we only maintain that this assertion cannot be proved. This calls for a further explanation, and we shall give it by showing the reason we cannot deduce from the first five axioms that there exist more than four points.

What is the meaning of the statement: "It can be proved that there exist more than four points"? This statement means that the following can logically be deduced: If a system consists of two kinds of things, called "points" and "lines," satisfying the five axioms, then it has more than four "points." It turns out, however, that we can easily show a system of "points" and "lines" which satisfies the five axioms but which contains just four points.

We shall denote by "points" four (arbitrary) things, indicated by the symbols

$$\square, \quad \bigcirc, \quad +, \quad -.$$

We denote by lines the six (arbitrary) things, indicated by

$$[\square\,\bigcirc], \quad [\square\,+], \quad [\square\,-], \quad [\bigcirc\,+], \quad [\bigcirc\,-], \quad [+\,-].$$

We shall agree to say that a line passes through a point if the symbol for the line contains the symbol for that point. Hence, to give an example, the line $[\square\,\bigcirc]$ passes through the point $\square$ and through the point $\bigcirc$.

The definitions of "parallel to" and "intersecting," given above, remain in force. Hence, the line $[\square\,\bigcirc]$ is parallel to the line $[+\,-]$, because these lines do not both pass through one of the points $\square$, $\bigcirc$, $+$, or $-$. Line $[\square\,+]$ intersects line $[+\,-]$; their point of intersection is the point $+$.

We now proceed to show that this system of points and lines satisfies the five axioms.

1. Through two points, for example, □ and O, passes a line, [□ O].
2. There is no other line passing through these two points: the line [□ +], for example, does not pass through both □ and O.
3. There are three points, for example, ⊔, O, and +, such that there is not a line passing through all three of them. Every line passes through two points only.
4. Line [□ O], for example, does not pass through the point +, and there is a line, [+ −], passing through + and parallel to [□ O].
5. There is no other line passing through + and parallel to [□ O].

Although this system has only four points, it nevertheless satisfies the five axioms. Hence, from the assumption that a certain system satisfies the five axioms, it cannot logically be deduced that this system has more than four points. Therefore, the existence of more than four points cannot be proved.

This is the reason why it could not be proved that the points D, E, and F in Figure 7-10 are different points.

Remark The preceding discussion was accomplished without any reference to a figure. The example makes it clear that we can choose any objects we like for the concepts mentioned in the axioms. The reader may find it helpful to construct a geometric model for the example. See Exercises 7-11.

EXERCISES 7-11

1 Consider the following four-point geometry (see Figure 7-13). The points are 1, 2, 3, and 4, the vertices of the tetrahedron. The lines are the six edges of the tetrahedron. Edges that have a vertex in common will be called *intersecting lines*. Edges that are not intersecting will be called *parallel lines*.

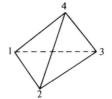

Figure 7-13

(a)   Complete the following table.

(1) Point	(2) Lines passing through point in column (1)	(3) Line	(4) Line parallel to line in column (3)	(5) Lines intersecting line in column (3)
1	(1, 2) (1, 3), (1, 4)	(1, 2)	(3, 4)	(1, 3), (1, 4), (2, 3), (2, 4)
2	. . .	(1, 3)	. . .	. . .
3	. . .	. . .	. . .	. . .
4	. . .	. . .	. . .	. . .
	. . .	. . .	. . .	. . .
	. . .	. . .	. . .	. . .

(b) Show that this geometry satisfies Axioms 1 through 5 of Section 7-10.

(c) Establish a one-to-one correspondence between the points and the lines of the four-point geometry of this problem and the points and lines of the four-point geometry in the text.

2 Consider the following four-point geometry (see Figure 7-14). The points are A, B, C, and D, the vertices of the quadrilateral. The lines are the sides and the diagonals of the quadrilateral. Lines that have a point in common will be called intersecting lines. Lines that are not intersecting will be called parallel lines.

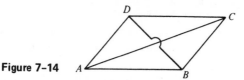

Figure 7–14

We assume that the diagonals do not intersect.

(a) Make a table similar to the one in Exercise 1(a).

(b) Show that this geometry satisfies Axioms 1 through 5 of Section 7-10.

(c) Establish a one-to-one correspondence between the points and lines of this four-point geometry and the points and lines of the four-point geometry in the text.

3 Consider the following system (see Figure 7-15). The points are P, Q, R, and S, the vertices of the quadrilateral. The lines are the sides of the quadrilateral, the diagonal PR, and the curve QS. Lines that have a point in common will be called intersecting lines. Lines that do not intersect will be called parallel. Show that the system is a four-point geometry satisfying Axioms 1 through 5 of Section 7-10.

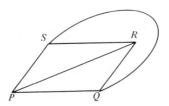

Figure 7–15

REFERENCES

For further discussions of Greek mathematics, see references 7 (Chapter 1) and 19, 20, and 21 (Chapter 3). Biographies of many Greek mathematicians may be found in encyclopedias and in the *Dictionary of Scientific Biography*.

Books on Archimedes and Diophantus are

[34] Dijksterhuis, E. J., *Archimedes*. Copenhagen: Munksgaard Ltd., 1956.

[35] Heath, T. L., *The Works of Archimedes with the Method of Archimedes*. New York: Dover Publications, Inc., no date (paperback).

[36] Heath, T. L., *Diophantos of Alexandria, A Study in the History of Greek Algebra*. New York: Dover Publications, Inc., 1964 (paperback).

For further data on some of the difficulties in the foundations of Euclidean geometry, see

[37] Moise, Edwin E., *Elementary Geometry from an Advanced Standpoint*. Reading, Mass.: Addison-Wesley Publishing Company, Inc., 1974.

[38] Prenowitz, Walter, and Meyer Jordan, *Basic Concepts of Geometry*. Lexington, Mass.: Xerox College Publishing, 1965.

There is an English translation of Hilbert's famous but concise book, and there is a high school textbook that is based on Hilbert's approach:

[39] Hilbert, David, *Foundations of Geometry*. Chicago: Open Court Publishing Company, 1902; La Salle, Ill.: Open Court Publishing Company, 1971.

[40] Brumfiel, C. F., R. E. Eicholz, and M. E. Shanks, *Geometry*. Reading, Mass.: Addison-Wesley Publishing Company, Inc., 1960.

There are several accounts of finite geometries. Additional references may be found in the second of the following two sources.

[41] Albert, A. A., "Finite Planes for the High School," *The Mathematics Teacher*, Vol. 55 (1962), pp. 165-169.

[42] Coxford, Arthur F., Jr., "Geometric Diversions: A 25-Point Geometry," *The Mathematics Teacher*, Vol. 57 (1964), pp. 561-564.

8

NUMERATION AND ARITHMETIC
AFTER THE GREEKS

8-1 ROMAN NUMERALS

The Greek ascendancy in Europe, Africa, and the Near East gave way to a succession of military conquests. Chronologically these were at the hands of the Romans, the Arabic-Moslem peoples, and, finally, Western Europeans. Cultures are not conquered or destroyed, however. They are modified and adopted by the conquerers. What is true for culture in general is also true for mathematics. In this chapter we shall discuss how the numeration systems of various cultures contributed to the development of our modern system.

Several symbols which are used in the modern version of Roman numerals represent changes from the earliest Roman numeration system. In the earliest system, the symbols were

$| = 1$, $V = 5$, $X = 10$, $L = 50$, $C = 100$, $|) = 500$, and $C|) = 1000$.

The symbol for 500 was part of the symbol for 1000. It became the letter "D" in later years, when printers saw a way of simplifying the complexity of their type. A similar simplification came in when M was introduced for 1000. There was a loss of uniformity here, however, because the original symbol for 1000 was easily made into

$CC|))$, for 10,000   and $CCC|)))$, for 100,000.

Various devices were adopted to represent large numbers with the newer symbols. A line over a set of Roman numerals multiplied the number they represented by 1000. Vertical lines on both sides multiplied the number by 100. Thus,

$$\overline{X} = 10,000, \qquad |X| = 1000,$$

while

$$|\overline{X}| = 10,000 \cdot 100 = 1,000,000.$$

The subtractive principle caused earlier Roman numerals, such as

$$|||| = 4 \qquad \text{and} \qquad LXXXX = 90$$

to be replaced by

$$IV\ (5-1) \qquad \text{and} \qquad XC\ (100-10).$$

Some other examples of these later Roman numerals are:

$$XIX = 19 \qquad\qquad CDXCI = 491$$
$$LIV = 54 \qquad\qquad MXV = 1015$$
$$CXLVII = 147 \qquad MCMLXXVI = 1976$$

The Romans made use of fractions with denominators that were multiples of 12. This use was related to their system of money and weights. The Roman *as* was a copper coin weighing 1 pound. It was equivalent to 12 *unciae*. The division of an *as* into 12 *unciae* resulted, at a later time, in the division of 1 pound (troy) into 12 ounces and a foot into 12 inches. The words "inch" and "ounce" come from *uncia*.

There are some clear connections between the Latin names of Roman numerals and English words. We have already mentioned descendants of the Latin *decem*, such as the English "decimal." The Latin word for 100, *centum*, was the forerunner of such English words as "cent," "century," and "centennial." The Latin word for 1000, *mille*, is the source of our "mile." Roman army marches were measured in *mille passuum*, 1 thousand paces; a Roman pace was two steps, approximately 5 feet. Thus, not only the word "mile" but also the approximate size of the mile (5280 feet) has descended to us from the Romans.

The Roman system of numeration was unsuitable for computation, for which purpose the *abacus* was used. The Roman numerals were mainly used for recording data and the results of calculations.

EXERCISES 8-1

1 Write the following addition problems in Roman numerals and then perform the addition.

(a) 2345   (b) 2345   (c) 327
       422   487   495
                                              601

2 Write the following subtraction problems in Roman numerals and then perform the subtraction.

(a) 2432   (b) 2432
       1321   1261

3 Repeat Exercises 1 and 2 by using Egyptian numerals.

4 See Exercises 1, 2, and 3. Discuss and compare the facts and principles that had to be understood to perform addition and subtraction in Roman, Egyptian, and modern notation.

5 Look up and report on
   (a) theories for the origins of the Roman symbols;
   (b) other ancient numeration systems, such as Minoan, Cretan, Etruscan, Phoenician, and Syrian (see reference 43, this chapter);
   (c) ancient and modern Hebrew, Chinese, and Japanese systems (see references 43, this chapter, and 23, Chapter 3).

Analyze each of the systems for the presence or absence of the principles we have listed here and in earlier chapters (base, place; repetitive, additive, multiplicative, and subtractive principles).

8-2 THE ABACUS AND TANGIBLE ARITHMETIC

The abacus, in its various forms, is in use in many parts of the world. In various places it is called by various names: *abacus, soroban, suan-pan, choty*. It was used by the ancient Egyptians, Greeks, and Romans as well as by the Arabs and medieval Western Europeans. Modern forms of the abacus are used in elementary schools as a visual aid to the teaching of place value, borrowing, and carrying.

The word *abacus* comes from a Greek word for a sand or dust table on which computations were done by making easily erased impressions in the sand. The Romans used for an abacus a metal plate or wooden board with grooves cut in it. Numbers were represented by little counters or stones, which were placed in the grooves. The Latin word for "stone," *calx*, led to our words "chalk" and "calcium"; its diminutive plural form, *calculi*, meant "little stones." This is the source of our words "calculate" and "calculus."

The *counting-board*, a later form of the abacus, was widely used after Roman times. The board was ruled as shown in Figures 8-1 and 8-2, where we have labeled it with Arabic numerals. The number represented in the left portion of Figure 8-1 is 2907; in the right portion the number 43 is shown. Note that there are never more than four counters on a line. When a 5, a 50, or a 500 is needed, a counter is placed between the appropriate lines. A line having no counter corresponds to a

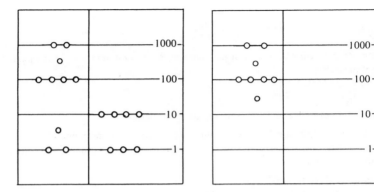

Figure 8-1   **Figure 8-2**

zero. To add the numbers 2907 and 43, push all the counters on the units line to the left part of that line, note that there are five of them, remove them and place ("carry") one in the space between the units line and the tens line. Now that there are two counters there, remove them and carry one to the tens line. Pushing all the counters on the tens line to the left produces 5. They are then removed and one is carried to the space between the tens and the hundreds line. Now the calculation has been completed because the counters have all been removed from the right portion of the board, no more than four counters remain on one line, and no more than one counter remains in each space between the lines. The sum is represented in Figure 8-2 and is 2950. If another number were to be added to the sum, it would be placed on the empty right portion of the board, and the process would be repeated. The explanation of subtraction and of the word "borrow," as used in subtraction, is left to the reader. (See Exercise 2 of Exercises 8-2.)

The English word "counter" comes from the practice of using a counting board as a table where business was transacted.

The Germans called a counting board a *Rechenbank*, or *Bank*, from which we get our words "bank" and "bankrupt." The latter referred to a merchant whose Bank had been broken as a symbol of his failure.

Many computing devices have been used since the invention of the abacus. These include Napier's bones, sector compasses, slide rules, calculators, and computers.

The term *tangible arithmetic* refers to the use of physical devices to record or compute with numbers. The ancient Peruvians kept local and state records on bundles of knotted cords each of which was called a *quipu*. Medieval merchants used an international system of finger symbols in their bargaining. *Tally sticks*, pieces of wood notched to record numbers, have been found from England to China. They were notched in various codes to record such things as quantities of milk or grain produced or of money loaned. Sticks that recorded loans were often split in the manner diagrammed in Figure 8-3. The larger piece with the handle was called the *stock*. The person lending the money kept it. He was called the *stockholder*. The borrower kept the other piece until the time came to settle up. The borrower and the lender then showed that their records agreed or *tallied* by fitting the parts together. The word *tally* itself comes from the same old French word meaning "to cut" as does our word *tailor* for "one who cuts cloth." Similarly, the word *score* for "twenty" comes from the use of a deeper cut or "score" to mark the twentieth notch or to record a group of 20 on a tally stick.

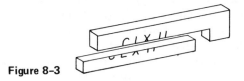

Figure 8-3

EXERCISES 8-2

1 Make a counting board and perform the following additions:
   (a) 5280   (b) 6927   (c) 4789   (d) 1234
       2316   2436   9678   5678
                                                 8765

2 Use the counting board to perform the following subtractions:
   (a) 468   (b) 5280   (c) 402
       321   2316   168

3 Read and report about the nature of tally sticks and about their use in different countries. See reference 44 (this chapter).

4 Reference 44 has pictures of *jetons* and *Rechenpfennige*. At one time the ability to use *jetons* was believed to make a girl a desirable wife. Report on these items of tangible arithmetic, where and how they were used, and what they had to do with marriage.

5 Find an account of a *quipu* in an encyclopedia or in reference 43 or 44 (this chapter). Make a model of one and prepare a report on how and when it was used.

6 For a series of short articles on tangible arithmetic, see reference 45 (this chapter). Carry out the following projects.
   (a) Make a model of Napier's bones and use them to perform the multiplication 234 · 58.
   (b) Make a model of Genaille's rods and use them to perform the multiplication 58 · 234.
   (c) Show how to use the sector compasses to solve the equation $\dfrac{x}{4} = \dfrac{3}{5}$.

(d) Show how to make the finger symbols for the natural numbers less than 100.

7 Look up the various forms of the *abacus*, the *soroban*, the *suan-pan*, and the *choty* (or *s'choty*) in an encyclopedia (or other source) and describe their differences. Which is the simplest? Why?

8-3 THE HINDU-ARABIC NUMERALS

The system of numeration used in most parts of the modern world is the Hindu-Arabic system. As we shall see, the complete system was developed much later than the systems that we have studied so far.

In prehistoric times the Hindus had settled in India. The earliest Hindu numerals, the *Brahmi numerals*, appear on inscriptions dating back to the third century B.C. The system resembled that of the Greeks; there were not only symbols for 1 through 9 but also for 10, 20, 30, ..., 90, 100, 200, and so on. Since there were separate symbols for 10, 20, and so on, it appears that positional notation of numbers was not yet known, no more so than with the Greeks. Thus, a symbol for "zero" was not needed and consequently not used.

The first 9 Brahmi numerals looked as follows:

$$- \quad = \quad \equiv \quad \text{Ψ} \quad \text{Γ} \quad \text{6} \quad 7 \quad \text{5} \quad ?$$

1　2　3　4　5　6　7　8　9.

The place-value system with 10 symbols, including a zero, may have been developed as early as A.D. 500, the time of the mathematician *Aryabhata*, but some scholars date the common use of the system after A.D. 700. Whether the Hindus themselves were the inventors of the system cannot be stated with certainty. They had undoubtedly come into contact with Greek culture and had taken cognizance of the work of the astronomer Ptolemy, who used a zero and calculated with sexagesimal fractions. At that time, their notation had the following form:

$$\text{٦} \quad \text{٢} \quad \text{3} \quad \text{४} \quad \text{५} \quad \text{(} \quad \text{7} \quad \text{८} \quad \text{9} \quad \text{o}$$

1　2　3　4　5　6　7　8　9　0.

These numerals are called *Indian numerals*. It can be assumed that they proceeded from the Brahmi numerals.

Two types of symbols evolved from the Indian numerals, the *West Arabic* or *Gobar numerals*:

$$\text{1} \quad \text{2} \quad \text{3} \quad \text{٤} \quad \text{5} \quad \text{6} \quad \text{7} \quad \text{8} \quad \text{9}$$

1　2　3　4　5　6　7　8　9

and the *East Arabic* numerals:

$$| \quad P \quad P \quad P \quad \Diamond \quad \Upsilon \quad \vee \quad \wedge \quad 9 \quad \bullet$$
$$1 \quad 2 \quad 3 \quad 4 \quad 5 \quad 6 \quad 7 \quad 8 \quad 9 \quad 0.$$

The first of these clearly reveal our modern numerals. The East Arabic numerals are still used in Turkey and the Arab countries. The question now arises: How did the people in the Western world come to use the Gobar numerals?

After A.D. 622, following *Mohammed's* flight to Medina and his return, the Arabs started their wars of conquest. They not only succeeded in subjugating all of North Africa, but they also founded one huge empire, which extended from India to Spain. Because the conquerors were relatively tolerant in this empire, the arts and sciences were intensively studied by non-Arabs as well as Arabs. Baghdad, founded in 766 by Caliph Al'Mansur, became a center for science and a depository for manuscripts. It was visited by scholars from all over the world. Much of our knowledge of Greek and Indian learning was preserved in Arabic translations made by scholars working there.

It was in this vast Arabic empire that the Arabic numerals came into common use. They became known especially by a book of the Arabian mathematician *Mohammed ibn Musa al-Khowarizmi* (Mohammed, the son of Moses of Khowarizm), written in the year 820 under the title *Al-jebr w'al-muqabala*. We have written these names in detail because part of the author's name has given rise to our word "algorithm," and because the title of the book was corrupted to "algebra et almucabala" and finally abbreviated to "algebra."

Al-Khowarizmi also wrote a booklet on Indian arithmetic in which he discusses the zero. He says: "When after subtracting, nothing is left, write a small circle, lest the place remains empty. The small circle must occupy the place, lest there will be fewer places and, for instance, the second is taken for the first." This shows that in the positional system the zero was a problem that required special explanation. That on the one hand zero represents "nothing" but on the other hand could change the number represented by the nonzero digits of a numeral was understood only with difficulty.

The Arabs introduced the Gobar numerals into Spain, from whence they gradually became known in Western Europe. The first person who attempted to spread the use of the Gobar numerals in Western Europe was *Gerbert* (later Pope Sylvester II) in the tenth century. During one of his journeys in Spain he became acquainted with the Arabic numerals. He realized the advantage of this notation and wrote a booklet in which he promoted the use of the new symbols and described how to calculate with them on the abacus. The counters on the abacus were to be replaced by disks (*apices*) on which Gobar numerals were written. Now, in order to represent, for example, the number 314 on the abacus, one no longer had to put 3, 1, and 4 counters on the columns of the hundreds, the tens, and the units, respectively; it would suffice to put in each column one disk with one of the numbers 3, 1, and 4 written on it.

However, if one supposes that the Arabic numerals now rapidly came into use, he or she is mistaken. The reason they did not is that calculating on the abacus with apices was not easier in every respect than calculating with counters. In the latter method addition and subtraction could be performed by simply adding or removing counters, and so it was not necessary to know the results of easy additions and subtractions by heart. This way of calculating was impossible with apices, that is, with the Gobar numerals. The consequence of this difficulty of calculating with apices was that the use of the abacus in Gerbert's way did not become popular after all and therefore did not promote the use of the Gobar numerals.

There were, no doubt, enlightened persons who recognized that the Gobar numerals would be highly convenient for calculation on paper. We should bear in mind, however, that writing material was scarce. The use of the abacus required only the recording of results, so little paper was needed for calculation.

We shall not be far from the truth, therefore, if we consider the scarcity of writing material as one of the causes of the continued use of the abacus. For the purpose of recording results, the old and dependable Roman numerals were sufficient. All in all, we need not be surprised that, for the time being, the Gobar numerals did not gain acceptance.

On this point it is interesting to recall that while Gerbert was still advocating, without much success, the use of the Gobar numerals (with the abacus), they were in common use in Arabia. Perhaps one of the reasons is that there was a paper mill in Baghdad as early as 794, whereas it was not until 1154 that there was one in Western Europe (in Spain, in fact).

After the year 1000 a struggle arose in Western Europe between the two methods of calculation, that of the "abacists" and that of the "algorists." The names speak for themselves. The struggle was eventually won by the algorists. Italy, the center of commerce at the time, was first in recognizing the advantages of calculating in writing; one of the advantages was that calculations could be checked afterward. In 1202 a book appeared, *Liber Abaci*, by *Leonardo of Pisa* (also called *Fibonacci*, "son of a good man"). This book had much influence, not only on mathematicians but also on merchants, and stimulated the diffusion of the West Arabic numerals. It is true that the Florentine municipality made attempts as late as 1299 to retard progress by putting a ban on the new symbols, under the pretext that accounts could be easily falsified; but malicious tongues had it that the town officials who had to verify the books did not themselves know the new method of calculation and tried to hinder it. But the advantages were too conspicuous, and so the method penetrated into Germany (1200), France (1275), and England (1300). *It was the middle of the sixteenth century, however, before the matter was decided definitely in favor of the West Arabic numerals.*

As long as people were obliged to write everything by hand, the form of the figures changed in the course of years. Since the invention of the art of printing, however, they have virtually maintained their appearance.

It is interesting to note that the word "cipher" was derived from the Arabic *al-sifr*, which means zero. In later Latin this became *cephirum*, and via the Italian *zevero* it became "zero." Al-sifr had been derived from the Hindu word *sunya* meaning "empty" or "void."

Kushyar Ibn Labban (971-1029), an Arab writing about Hindu numerals, typifies the writers of his time in his use of base 10 numerals for whole numbers but in writing fractions sexagesimally. He referred to whole numbers as degrees, $\frac{1}{60}$ as a "minute," $\frac{1}{60^2}$ as a "second," $\frac{1}{60^3}$ as a "third," and so on, and indicated them by marks, $°, ', '', '''$, and so on. He then stated: "The results of the multiplication of degrees by degrees are degrees, and degrees by fractions are those fractions, as degrees by minutes are minutes, and by seconds are seconds. For fractions by fractions, it is a gathering of the marks, as minutes by seconds are thirds, because it is 1 plus 2 . . . ." A modern reader sees in the latter statement a foreshadowing of both the use of exponents and of the rules for modern multiplication with decimal fractions. Thus, 3 minutes times 2 seconds would have been written as $3'$ times $2''$, and the result would have been $6'''$, or six "thirds." Today, we would say:

$$\frac{3}{60^1} \cdot \frac{2}{60^2} = \frac{6}{60^3},$$

where the exponent 3 is obtained as the sum of 1 and 2. In the elementary school, we teach the corresponding algorithm for $0.2 \times 0.03 = 0.006$ by telling children to multiply the numbers 2 and 3 and then "point off" the number of decimal places equal to the sum of the numbers of decimal places in the two factors.

Ibn Labban's work illustrates two additional remarkable facts: the persistence of sexagesimal fractions from the Babylonians until today, and the slowness of the adoption of the analogous decimal fractions.

It was not until about 1600 that the idea of writing fractions in the form of decimals was promoted in Western Europe. The Dutch mathematician *Simon Stevin* (1548-1620) was the first to throw clear light upon the advantages of the decimal notation. In a pamphlet, *De Thiende* (The Dime), he advocated the use of decimal fractions.

He urged that governments adopt the decimal system and also decimal coins, weights, and measures. This did not happen on a large scale, however, until the French Revolution.

8-4   AN EARLY AMERICAN PLACE-VALUE NUMERATION SYSTEM

The *Mayas* of Central America and Mexico had developed a remarkable civilization long before Columbus came to the Western hemisphere. Their numeration system was a base 20 place-value system with a zero. They wrote their numerals in a column, the units position being the lowest. Only three symbols were used: a dot for 1, a short line for 5, and an egg-shaped symbol for zero.

EXAMPLES

$$\begin{array}{l} \underline{\overset{\bullet}{\equiv}} = 16 = 3\cdot 5 + 1 \end{array}$$

$$\underline{\bullet\bullet\bullet\bullet} = 9 = 5 + 4$$

$$\begin{array}{l} \overset{\bullet\bullet\bullet\bullet}{\underline{\bullet}} \\ \underline{\equiv} \end{array} = 196 = 3\cdot 5 + 1 + (5 + 4) \cdot 20$$

$$\begin{array}{l} \overset{\bullet}{\equiv} \\ \bigotimes \end{array} = 320 = 0 + (3\cdot 5 + 1) \cdot 20$$

The Mayas departed from a pure base 20 system by letting the symbols in the third position (from the bottom) represent the number of 360's $(360 = 18 \cdot 20)$ rather than 400's $(400 = 20^2)$. The reason may be that their year consisted of 360 days.

EXAMPLES

$$\begin{array}{l} \overset{\bullet\bullet}{\underline{\bullet\bullet\bullet}} \\ \overset{\bullet}{} \end{array} = 2586 = (5 + 1) + 3 \cdot 20 + (5 + 2) \cdot 360$$

$$\begin{array}{l} \overset{\bullet}{\bigotimes} \\ \bullet\bullet\bullet\bullet \end{array} = 320 = (5 + 4) + 0 \cdot 20 + (5 + 1) \cdot 360$$

Symbols in the fourth position represented the number of 7200's $(7200 = 18 \cdot 20^2)$, and so on.

EXERCISES 8-4

1 Write in Mayan notation:
  (a) 19
  (b) 412
  (c) 52,800
  (d) 12,489,781 (the largest number found in a Mayan codex).

2 What numbers are represented by the following numerals?

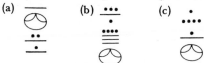

3 (a) Find the largest number that can be represented in Mayan notation by using three positions.
  (b) Find the smallest number that can be represented by using four positions.
  (c) Is there any integer that could not be represented?

4 Repeat Exercise 3 for a system in which the second position represents 20, the third position represents $18 \cdot 20$, and the fourth position represents 20^3.

8-5 LATER DEVELOPMENTS IN POSITIONAL NOTATION

Some of the properties of a positional system of numeration were mentioned in Chapter 2 in connection with Babylonian numerals. In the present section we shall discuss positional systems in more detail.

Some requirements for a theory of positional numeration systems are:

1. The notion that any counting number greater than 1 can be used as a *base*.

2. A set of distinct *symbols* (digits), including a zero, for the whole numbers less than the base.

3. The *multiplicative place-value* principle that a digit written in a particular place represents the product of the number represented by the digit and the power of the base corresponding to the place of the digit.

4. The *additive* principle that the number represented by a given numeral is the sum of the products in (3).

5. The idea of extending the system to include fractions.

6. The idea of using a symbol (a period) to separate the whole-number part and the fractional part of a numeral.

The German mathematician *Gottfried Wilhelm Leibniz* (1646-1716) was the first to give a generalized treatment of positional numeration systems. It has recently been discovered that the Englishman *Thomas Harriot* (1560-1621) had discussed these ideas earlier and written about them in an unpublished journal.

Leibniz was particularly intrigued with the idea of the binary system, which uses 2 as its base. In this system the successive places represent powers of 2. Thus,

$$1011_2$$

represents

$$1 \cdot 2^3 + 0 \cdot 2^2 + 1 \cdot 2^1 + 1,$$

which is equal to 11. The table shows the binary representation of the integers 1 to 7. It illustrates the fact that in the binary system only two digit symbols are needed, 0 and 1. Any integer can be represented by using only ones and zeros.

Binary	Decimal
1	1
1 0	2
1 1	3
1 0 0	4
1 0 1	5
1 1 0	6
1 1 1	7

Leibniz, who was a philosopher and a theologian as well as a mathematician, saw an analogy between the binary system and the creation of the universe as related in Genesis. He associated the fact that all whole numbers can be represented in the binary system by means of only two different digits, with the story that the universe was created out of a void (0) by God (1). Figure 8-4 pictures the title page of a book that appeared after Leibniz' death. The figure shows a commemorative coin or medallion which Leibniz urged his patron, the Duke of Braunschweig-Lüneburg, to have manufactured. Apparently, the coin was never struck. However, it is interesting to notice that it would have included our binary table, extended to 17, and examples of addition and multiplication on the left and right of the table, respectively.

We do not know what other applications of the binary system Leibniz had in mind, but he certainly was not thinking of the modern computer, in which the system is applied.

The binary system has often been used in the solution of puzzles and in the explanation or design of games (for example, Nim). See Exercise 1 of Exercises 8-5 for a mind-reading trick.

Figure 8-4 Title page of a book in the Charles Patterson Van Pelt Library of the University of Pennsylvania.

EXERCISES 8-5

1 The following is a mind-reading trick.
   (a) Make a set of cards such as those shown here.
   (b) Ask somebody to think of one of the numbers 1 through 15.
   (c) Show him the cards one by one and ask him each time if the number he has chosen is on the card.
   (d) Add the first numbers of the cards for which his answer is "yes."
   Then the sum is the chosen number.

1	9
3	11
5	13
7	15

2	10
3	11
6	14
7	15

4	12
5	13
6	14
7	15

8	12
9	13
10	14
11	15

2 Explain why the trick in Exercise 1 works. (Hint: Write the numbers on each card in binary notation.)

3 (a) Design a mind-reading trick, such as the one in Exercise 1, for the numbers 1 through 31. (Hint: Five cards will be needed, and the first numeral on the fifth card will be 16.)
   (b) Explain why the trick in (a) works.

4 (a) Look up the rules of the game of Nim.
   (b) Design a winning strategy for Nim and explain why it works. (Hint: The binary system is useful in the explanation of a winning strategy.)

5 Read and report on a biography of Leibniz.

6 Read and report on a biography of Harriot.

8-6   CONVERSIONS BETWEEN NUMERATION SYSTEMS

Any natural number greater than 1 may be used as the base for a positional numeration system. Once the base, b, is chosen, b different digits are needed: a symbol for zero and symbols for the $b - 1$ natural numbers less than b. The base used in the representation of a number is indicated by a subscript, if necessary. If no base is indicated, we shall assume that the base is ten.

In the following example, the first 14 natural numbers are represented in four systems, each with a different base.

Base ten:

1   2   3   4   5   6   7   8   9   10   11   12   13   14

Base 7:

1   2   3   4   5   6   10   11   12   13   14   15   16   20

Base 5:

1   2   3   4   10   11   12   13   14   20   21   22   23   24

Base 2:

1 10 11 100 101 110 111 1000 1001 1010 1011 1100 1101 1110.

If the base is greater then ten, we must invent some symbols for the numbers greater than 9 and less than the base. Thus, in a duodecimal (base 12) numeration system we need new single symbols for 10 and 11. The Duodecimal Society's symbol and name for the number ten are "X" and "dek" and for the number eleven "E" and "elf." The first 14 natural numbers are represented in this system as follows:

1   2   3   4   5   6   7   8   9   X   E   10   11   12.

Some examples will show how to convert a numeral from one system to another.

EXAMPLE 1   Convert 135_7 to a base ten numeral.

$$135_7 = 1 \cdot 7^2 + 3 \cdot 7 + 5$$
$$= 49 + 21 + 5 = 75.$$

EXAMPLE 2   Convert $2XE98_{12}$ to a base ten numeral.

$$2XE98_{12} = 2 \cdot 12^4 + 10 \cdot 12^3 + 11 \cdot 12^2 + 9 \cdot 12 + 8$$
$$= 60,452.$$

A second way to represent numbers in a base 12 system would be to use our base ten symbols for the numbers ten and eleven, that is, use "10" and "11" as new "digits." In this case it would be necessary to separate the digits of the duodecimal numeral with commas, as we did in dealing with the sexagesimal system (see Section 2-2). For instance, the number in Example 2,

$$2XE98_{12},$$

would then be written in the form

$$(2, 10, 11, 9, 8)_{12}.$$

The following examples show how a number expressed in base ten is converted to another base.

EXAMPLE 3   Convert 198 to base 5.

Consider the following powers of 5:

$$5^1 = 5, \qquad 5^2 = 25, \qquad 5^3 = 125, \qquad 5^4 = 625.$$

The largest power of 5 contained in 198 is 125. It is contained one time. Subtraction gives

$$198 - 1 \cdot 125 = 73.$$

The largest power of 5 contained in 73 is 25. It is contained two times. Subtraction gives

$$73 - 2 \cdot 25 = 23.$$

The largest power of 5 contained in 23 is 5. It is contained four times. Subtraction gives

$$23 - 4 \cdot 5 = 3.$$

The remainder 3 is less than 5. Hence,

$$198 = 1 \cdot 125 + 2 \cdot 25 + 4 \cdot 5 + 3$$
$$= 1 \cdot 5^3 + 2 \cdot 5^2 + 4 \cdot 5 + 3$$
$$= 1243_5.$$

EXAMPLE 4   Convert 135 to base 7.

Some powers of 7 are

$$7^1 = 7, \qquad 7^2 = 49, \qquad 7^3 = 343.$$

$$\begin{aligned}135 &= 2 \cdot 49 + 37 \\ &= 2 \cdot 49 + 5 \cdot 7 + 2 \\ &= 2 \cdot 7^2 + 5 \cdot 7 + 2 \\ &= 252_7.\end{aligned}$$

EXAMPLE 5   Convert 735 to base 9.

$$\begin{aligned}735 &= 1 \cdot 729 + 6 \\ &= 1 \cdot 9^3 + 0 \cdot 9^2 + 0 \cdot 9 + 6 \\ &= 1006_9.\end{aligned}$$

EXAMPLE 6   Convert 5938 to base 12.

$$\begin{aligned}5938 &= 3 \cdot 1728 + 754 \\ &= 3 \cdot 1728 + 5 \cdot 144 + 34 \\ &= 3 \cdot 1728 + 5 \cdot 144 + 2 \cdot 12 + 10 \\ &= 3 \cdot 12^3 + 5 \cdot 12^2 + 2 \cdot 12 + 10 \\ &= 352X_{12}.\end{aligned}$$

EXERCISES 8-6

1 Write in base ten
   (a) 5_8 (b) 10_8
   (c) 25_8 (d) 100_8
   (e) 574_8

2 Write in base 5:
   (a) 14   (b) 15
   (c) 16   (d) 24
   (e) 25   (f) 26

3 Write in base 5:
   (a) 2378
   (b) 750
   (c) 7843

4 Write in base 6:
   (a) 5   (b) 6
   (c) 24   (d) 36
   (e) 40   (f) 125
   (g) 328   (h) 1296

5 Write in base ten:
   (a) $39XE_{12}$ (b) 100101_2

6 Write 3230 in base 12.

7 Write in base 2:
   (a) 5   (b) 15
   (c) 22   (d) 32
   (e) 81

8 Write 234_5 in base 8.

9 Find the base (b) in each of the following:
   (a) $48 = 53_b$ (b) $93 = 233_b$

8-7 ADDITION AND SUBTRACTION ALGORITHMS IN NONDECIMAL BASES

The operations of addition and subtraction in nondecimal bases are performed in the same way as in base ten. It will be helpful to have addition tables, such as the one for base 6 shown below. One "carries" and "borrows" sixes in addition and subtraction as one carries and borrows tens when calculating in base ten.

Base 6:

+	0	1	2	3	4	5
0	0	1	2	3	4	5
1	1	2	3	4	5	10
2	2	3	4	5	10	11
3	3	4	5	10	11	12
4	4	5	10	11	12	13
5	5	10	11	12	13	14

EXAMPLE   Base 6:

```
    534          235
+   423       -   41
  1401          154
```

EXERCISES 8-7

1 Add in base 6:
   (a) 532   (b) 245   (c) 3542   (d) 413
       123   431   1134   524
                                                          123

2 Add in base 8:
   (a) 632   (b) 675   (c) 4562   (d) 276
       553   454   1456   142
                                                          614

3 Subtract in base 5:
   (a) 423   (b) 421   (c) 4310   (d) 34012
        221   234   3404   23243

4 Subtract in base 7:
   (a) 534   (b) 4201   (c) 54321   (d) 40001
        425   3322   12345   15362

8-8 MULTIPLICATION ALGORITHMS
IN NONDECIMAL BASES

As soon as we have the multiplication table for a certain base at our disposal, we can find products nearly as easily as in our system. This will be shown by multiplying 138 and 62 in three numeration systems.

Base ten:

X	0	1	2	3	4	5	6	7	8	9
0	0	0	0	0	0	0	0	0	0	0
1	0	1	2	3	4	5	6	7	8	9
2	0	2	4	6	8	10	12	14	16	18
3	0	3	6	9	12	15	18	21	24	27
4	0	4	8	12	16	20	24	28	32	36
5	0	5	10	15	20	25	30	35	40	45
6	0	6	12	18	24	30	36	42	48	54
7	0	7	14	21	28	35	42	49	56	63
8	0	8	16	24	32	40	48	56	64	72
9	0	9	18	27	36	45	54	63	72	81

Multiplication
```
  138
   62
  276
 828
 8556
```

Base 7:

X	0	1	2	3	4	5	6
0	0	0	0	0	0	0	0
1	0	1	2	3	4	5	6
2	0	2	4	6	11	13	15
3	0	3	6	12	15	21	24
4	0	4	11	15	22	26	33
5	0	5	13	21	26	34	42
6	0	6	15	24	33	42	51

Multiplication
```
  255
  116
 2262
 255
255
33642
```

Base 2:

X	0	1
0	0	0
1	0	1

Multiplication
```
      10001010
        111110
     100010100
     10001010
    10001010
   10001010
  10001010
10000101101100
```

EXERCISES 8-8

1 Draw up the addition and multiplication tables for the base 4 system.

2 Write 14 and 30 in the base 4 system; calculate their product with the aid of the tables of Exercise 1.

3 Multiply in base 4:

(a) 32123
          23

(b) 1212
         123

(c) 2130
        3121

4 Multiply in base 7 (first construct the multiplication table):

(a) 231
       46

(b) 314
       123

(c) 5432
       1653

5 Multiply in base 2:

(a) 10110
       1001

(b) 10110
       10111

(c) 110110
       101111

6 Divide in base 4:

(a) 21$\overline{)3303}$

(b) 23$\overline{)13233}$

(c) 223$\overline{)102021}$

7. Divide in base 7:

(a) 23$\overline{)16426}$

(b) 65$\overline{)34326}$

(c) 111$\overline{)41625}$

8-9 FRACTIONS, RATIONAL NUMBERS, AND PLACE-VALUE NUMERATION

The Babylonian system for writing fractions persisted long after the end of the Babylonian civilization. We shall therefore describe this system more fully. In doing so, we shall also discuss some general principles related to the representation of rational numbers as place-value fractions. When writing fractions, the Babylonians used place-value and base 60 in the same way that we use place-value and base ten. We shall use the modern notation for expressing sexagesimal fractions (see page 45).

The following example shows some conversions of fractions written in base ten to sexagesimal fractions.

EXAMPLE 1

Base ten representation	Sexagesimal fractions
$\dfrac{1}{12} = \dfrac{5}{60}$	0;5
$\dfrac{3}{5} = \dfrac{36}{60}$	0;36
$\dfrac{1}{3} = \dfrac{20}{60}$	0;20
$\dfrac{1}{6} = \dfrac{10}{60}$	0;10

The above conversions are simple because the denominators of the base ten fractions are factors of 60. If the denominator is not a factor of 60 but a product of factors of 60, the conversion is only a little more difficult.

EXAMPLE 2 Write $\frac{1}{9}$ as a sexagesimal fraction.

$$\frac{1}{9} = \frac{60 \cdot \frac{1}{9}}{60} = \frac{6\frac{2}{3}}{60} = \frac{6}{60} + \frac{\frac{2}{3}}{60} = \frac{6}{60} + \frac{60 \cdot \frac{2}{3}}{60^2}$$

$$= \frac{6}{60} + \frac{40}{60^2} = 0;6,40 \quad \text{(sexagesimal)}.$$

The following example shows what happens if we apply the same procedure to a fraction with a denominator containing a factor that is not a factor of 60.

EXAMPLE 3 Write $\frac{1}{7}$ as a sexagesimal fraction.

$$\frac{1}{7} = \frac{60 \cdot \frac{1}{7}}{60} = \frac{8 + \frac{4}{7}}{60} = \frac{8}{60} + \frac{\frac{4}{7}}{60} = \frac{8}{60} + \frac{60 \cdot \frac{4}{7}}{60^2} = \frac{8}{60} + \frac{34 + \frac{2}{7}}{60^2}$$

$$= \frac{8}{60} + \frac{34}{60^2} + \frac{60 \cdot \frac{2}{7}}{60^3} = \frac{8}{60} + \frac{34}{60^2} + \frac{17 + \frac{1}{7}}{60^3} = \frac{8}{60} + \frac{34}{60^2} + \frac{17}{60^3} + \frac{\frac{1}{7}}{60^3}.$$

If we were to continue, the last fraction would be converted as follows:

$$\frac{\frac{1}{7}}{60^3} = \frac{60 \cdot \frac{1}{7}}{60^4} = \frac{\frac{60}{7}}{60^4} = \frac{8}{60^4} + \frac{\frac{4}{7}}{60^3},$$

which leads to the same sequence of numerators, 8, 34, 17, as we obtained before. Thus, the representation of $\frac{1}{7}$ is as follows:

$$\frac{1}{7} = \frac{8}{60} + \frac{34}{60^2} + \frac{17}{60^3} + \frac{8}{60^4} + \frac{34}{60^5} + \frac{17}{60^6} + \frac{8}{60^7} + \cdots,$$

or $\frac{1}{7} = 0; 8, 34, 17, 8, 34, 17, \ldots,$

for which we write

$$\frac{1}{7} = 0; \overline{8, 34, 17}.$$

This is a repeating sexagesimal fraction, similar to the repeating decimal fractions with which we are familiar, such as $\frac{1}{3} = 0.333 \ldots = 0.\overline{3}$ and $\frac{4}{37} = 0.108108 \ldots = 0.\overline{108}$. We can condense the procedure shown in Example 3 into the following algorithm. Each stage of our process involved:

1. Multiplying the numerator $(1, 4, 2)$ of a fraction $(\frac{1}{7}, \frac{4}{7}, \frac{2}{7})$ by 60.
2. Dividing the product by the denominator (7).
3. Using the remainder $(4, 2, 1)$ as the numerator of a new fraction $(\frac{4}{7}, \frac{2}{7}, \frac{1}{7})$.

The quotients $(8, 34, 17)$ become the numerators of fractions the denominators of which are the successive powers of 60. These numerators are the "digits" of our sexagesimal-fraction representation.

The following scheme connects these steps:

$$
\begin{array}{l}
1 \\
\times\ 60 \\
\hline
7\lfloor 60 \\
\quad 8\ \text{rem.}\quad 4 \\
\qquad \times\ 60 \\
\qquad \hline
\qquad 7\lfloor 240 \\
\qquad\quad 34\ \text{rem. } 2 \\
\qquad\qquad \times\ 60 \\
\qquad\qquad \hline
\qquad\qquad 7\lfloor 120 \\
\qquad\qquad\quad 17\ \text{rem.}\quad 1 \\
\qquad\qquad\qquad \times\ 60 \\
\qquad\qquad\qquad \hline
\qquad\qquad\qquad 7\lfloor 60 \\
\qquad\qquad\qquad\quad 8\ \text{rem. } 4 \qquad \text{etc.}
\end{array}
$$

or
$$\frac{1}{7} = 0;\overline{8, 34, 17}.$$

The same process can be used to find a place-value representation of a fraction in any base. In base ten this process is the idea behind the usual division algorithm for converting a common fraction to a decimal fraction. To illustrate this, let us begin the conversion of $\frac{1}{7}$ to a decimal fraction, using the basic procedure, and then derive the algorithm from it.

$$
\begin{aligned}
\frac{1}{7} &= \frac{10 \cdot \frac{1}{7}}{10} = \frac{1 + \frac{3}{7}}{10} \\[2mm]
&= \frac{1}{10} + \frac{\frac{3}{7}}{10} = \frac{1}{10} + \frac{10 \cdot \frac{3}{7}}{10^2} = \frac{1}{10} + \frac{\frac{30}{7}}{10^2} \\[2mm]
&= \frac{1}{10} + \frac{4 + \frac{2}{7}}{10^2} = \frac{1}{10} + \frac{4}{10^2} + \frac{\frac{2}{7}}{10^2} = \frac{1}{10} + \frac{4}{10^2} + \frac{10 \cdot \frac{2}{7}}{10^3}
\end{aligned}
$$

$$= \frac{1}{10} + \frac{4}{10^2} + \frac{\frac{20}{7}}{10^3} = \frac{1}{10} + \frac{4}{10^2} + \frac{2 + \frac{6}{7}}{10^3}$$

$$= \frac{1}{10} + \frac{4}{10^2} + \frac{2}{10^3} + \frac{\frac{6}{7}}{10^3}$$

$$\vdots$$

Note that at each stage our steps were:

1. Multiplying the numerator $(1, 3, 2, \ldots)$ of a fraction $(\frac{1}{7}, \frac{3}{7}, \frac{2}{7}, \ldots)$ by 10.
2. Dividing the product by the denominator (7).
3. Using the remainder $(3, 2, 6, \ldots)$ as the numerator of a new fraction $(\frac{3}{7}, \frac{2}{7}, \frac{6}{7}, \ldots)$.

The quotients $(1, 4, 2, \ldots)$ become the numerators of fractions the denominators of which are the successive powers of 10. The numerators are the digits of our decimal-fraction representation. This procedure for writing $\frac{1}{7}$ as a decimal fraction can be condensed into a scheme which exactly parallels the previous scheme for writing $\frac{1}{7}$ as a sexagesimal fraction. Thus,

$$
\begin{array}{l}
\quad 1 \\
\times\ 10 \\
\hline
7\,\lfloor 10 \\
\quad 1 \text{ rem. } 3 \\
\qquad \times\ 10 \\
\qquad\hline
\quad 7\,\lfloor 30 \\
\qquad 4 \text{ rem. } 2 \\
\qquad\qquad \times\ 10 \\
\qquad\qquad\hline
\qquad 7\,\lfloor 20 \\
\qquad\qquad 2 \text{ rem. } 6 \\
\qquad\qquad\qquad \times\ 10 \\
\qquad\qquad\qquad\hline
\qquad\qquad 7\,\lfloor 60 \\
\qquad\qquad\qquad 8 \text{ rem. } 4 \\
\qquad\qquad\qquad\qquad \times\ 10 \\
\qquad\qquad\qquad\qquad\hline
\qquad\qquad\qquad 7\,\lfloor 40 \\
\qquad\qquad\qquad\qquad 5 \text{ rem. } 5 \\
\qquad\qquad\qquad\qquad\qquad \times\ 10 \\
\qquad\qquad\qquad\qquad\qquad\hline
\qquad\qquad\qquad\qquad 7\,\lfloor 50 \\
\qquad\qquad\qquad\qquad\qquad 7 \text{ rem. } 1 \qquad \text{etc.}
\end{array}
$$

At this stage, having reached a remainder, 1, which is the same as our original numerator, we see that the whole process would begin to repeat. Our result is:

$$\frac{1}{7} = \frac{1}{10} + \frac{4}{10^2} + \frac{2}{10^3} + \frac{8}{10^4} + \frac{5}{10^5} + \frac{7}{10^6} + \cdots$$

$$= 0.\overline{142857}$$

This algorithm amounts to the familiar division algorithm. Thus,

```
        0.142857
      7)1.000000
        7
        ──
        30
        28
        ──
        20
        14
        ──
        60
        56
        ──
        40
        35
        ──
        50
        49
        ──
        1
```

In this algorithm we affix zeros and bring them down. This obscures the real idea, that we are multiplying the numerators by 10 and dividing the results by the denominator. However, this affixing and bringing down of zeros makes the algorithm easier to remember because it reduces it to the familiar long-division algorithm.

At each step in the algorithms for converting $\frac{1}{7}$ to a decimal fraction we divided by 7. When one divides by 7, the possible remainders are 0, 1, 2, 3, 4, 5, and 6. If some remainder is zero, the process terminates (gives a zero in the quotient from then on). If no remainder is zero on or before the seventh division, one of the other possible remainders 1, 2, 3, 4, 5, or 6 must reappear. Hence, the number of digits in the *period* (the sequence of digits that repeats) cannot be greater than 6. This discussion suggests the following generalizations.

1. Every rational number can be represented in any base by a terminating or a repeating place-value fraction.
2. If the denominator of a fraction is a product of factors of the base, the place-value representation will terminate (repeat zeros).
3. The number of digits in the period of a repeating place-value fraction will be less than the denominator of the fraction that generates the place-value fraction.

EXERCISES 8-9

1 Write each of the following sexagesimal fractions as a decimal fraction.
   (a) 4,21;15   (b) 4,0,21;30

  (c) 4,21;0,30   (d) 0;45

  (e) 0;6   (f) 0;0,6

  (g) $261°15'$ (h) $4°15'30''45'''$

2 Do the following additions. Check your answers by converting to base ten fractions. (Hint: See the following example.)

Addition: 42;15,18

              21;30,45

             1,3;46, 3

Check: $42 + \dfrac{15}{60} + \dfrac{18}{60^2}$

           $21 + \dfrac{30}{60} + \dfrac{45}{60^2}$

           $63 + \dfrac{45}{60} + \dfrac{63}{60^2}$

and

$$63 + \frac{45}{60} + \frac{63}{60^2} = 63 + \frac{46}{60} + \frac{3}{60^2} = 1 \cdot 60 + 3 + \frac{46}{60} + \frac{3}{60^2} = 1,3;46,3.$$

  (a) 6;30   (b) 4;15,25

       2;15   1;44,50

  (c) 2,37;40   (d) $5°42'\ 3''4'''$

      1,23;30 $184°25'39''8'''$

3 Perform the following multiplications, retaining the sexagesimal representation of the numbers. Check your answers by converting to base ten fractions. (Hint: See the following example.)

Multiplication: $42°15'$

                    $2°30'$

              $1260'450''$

        $84°\ \ \ 30'$

        $84°1290'450'' = 105°37'30''.$

Check: $(42\tfrac{1}{4}) \times (2\tfrac{1}{2}) = \dfrac{169}{4} \times \dfrac{5}{2}$

                    $= \dfrac{845}{8}$

                    $= 105\tfrac{5}{8},$

which is equal to $105°37'30''.$

  (a) $(0;30) \times (0;15)$ (b) $(26°45') \times (5°20'')$

  (c) $(2;6,30) \times (12;0,45)$ (d) $(4°2'3'') \times (5°4'')$

4 Write as a sexagesimal fraction:

  (a) $\dfrac{1}{2}$ (b) $\dfrac{3}{4}$

  (c) $\dfrac{1}{5}$ (d) $\dfrac{5}{6}$

  (e) $\dfrac{1}{10}$ (f) $\dfrac{7}{12}$

(g) $\dfrac{8}{15}$ (h) $\dfrac{1}{20}$

(i) $\dfrac{13}{30}$ (j) $\dfrac{7}{60}$

5 Write as a sexagesimal fraction:

(a) $\dfrac{1}{8}$ (b) $\dfrac{1}{16}$

(c) $\dfrac{1}{18}$ (d) $\dfrac{1}{24}$

(e) $\dfrac{1}{25}$ (f) $\dfrac{1}{27}$

(g) $\dfrac{1}{32}$ (h) $\dfrac{1}{36}$

6 Write as a sexagesimal fraction:

(a) $\dfrac{1}{11}$

(b) $\dfrac{1}{13}$

(c) $\dfrac{1}{14}$

How many digits are there in the period of each? Compare the period of $\dfrac{1}{14}$ with that of $\dfrac{1}{7}$. Do you see any connection between them? If so, test your conjecture by making up some similar pairs of fractions and expanding them in sexagesimal notation.

7 Find a sexagesimal representation for

(a) $\dfrac{2}{5}$ (b) $\dfrac{2}{7}$

(c) $\dfrac{3}{5}$ (d) $\dfrac{3}{7}$

(e) $\dfrac{3}{14}$

Do you see any connections between these representations and your earlier expansions of $\dfrac{1}{5}$, $\dfrac{1}{7}$, and $\dfrac{1}{14}$?

8 Which unit fractions with denominators less than the indicated base can be written as terminating
   (a) decimal fractions?
   (b) sexagesimal fractions?
   (c) duodecimal (base 12) fractions?

9 Consider all the unit fractions with denominators less than 20, that is, the fractions $\dfrac{1}{2}$, $\dfrac{1}{3}$, ..., $\dfrac{1}{19}$. In which numeration system, the decimal, the duodecimal, or the sexagesimal, does a maximum number of these fractions have a terminating place-value representation?

10 List the advantages and disadvantages of each of the following numeration systems as compared with our decimal numeration system: the binary, the quinary (base 5), the duodecimal, the sexagesimal.

11 The duodecimal system was often treated in a separate chapter in English and early American arithmetics. It was useful because of its connection with weights and measures. As a result of this connection, the terminology and symbolism

of feet and inches were taken over and used even when no measurements of length were involved. The following examples are taken from Daboll's *Schoolmaster's Assistant* (mentioned in Exercise 3 of Exercises 1-10). Explain the first example and complete the others. Check your work, using common fractions with powers of 12 as denominators. (F. and I. stand for "feet" and "inches.")

	(a)	F.	I.		
multiply		7	3		
by		4	7		
		29	0		
		4	2	9	
product		33	2	9	

	(b)	F.	I.
		4	6
		5	8

	(c)	F.	I.
		9	7
		9	7

12 Figures 8-5(a), 8-5(b), and 8-5(c) show the title page and part of the section on duodecimals of another early American arithmetic.
   (a) See Figure 8-5(b). Write the answer to the first example of addition, by using feet and inches, and fractions of inches with powers of 12 as denominators.
   (b) See Figure 8-5(c). Do the subtraction problem shown as the second of the examples at the top of the figure.
   (c) See Figure 8-5(c). Explain the first worked example of multiplication.
   (d) See Figure 8-5(c). Do Example 2 in the "Multiplication of Duodecimals" section.

13 See the quotation of Kushyar Ibn Labban on page 229. Apply his idea to the multiplication of the sexagesimal fractions $3°2'15''$ and $2°4'5''$. Check your answer using common fractions with powers of 60 as denominators. Notice that Ibn Labban anticipated one of our laws of exponents.

14 Figure 8-6 shows a page from Ibn Labban's manuscript. The numerals in lines 8, 13, and 16 explain the steps in the process of adding two numbers. When the addition was done on a dust board, the second addend was retained throughout the process, but the first addend was erased step by step and replaced by the sum. Write these steps in modern symbols and explain the process. For the meaning of the numerals, see Table 8-1, page 259.

8-10   IRRATIONAL NUMBERS

In the previous section we showed that *every rational number can be represented by a terminating or a repeating decimal.* Thus,

$$\frac{1}{2} = 0.5$$

(a)

THE
American Tutor's Assistant
REVISED;

OR,

A COMPENDIOUS SYSTEM OF

PRACTICAL ARITHMETIC;

CONTAINING

THE SEVERAL RULES OF THAT USEFUL SCIENCE,
CONCISELY DEFINED, METHODICALLY ARRANGED, AND
FULLY EXEMPLIFIED.

THE WHOLE

PARTICULARLY ADAPTED TO THE EASY AND REGULAR
INSTRUCTION OF YOUTH IN OUR AMERICAN SCHOOLS.

Originally compiled by sundry Teachers in and near Philadelphia; now Revised, and an additional number
of Examples given in money of the
United States.

TO WHICH IS ADDED, A COURSE OF

Book-Keeping by Single Entry.

....

PHILADELPHIA:

Printed and sold by Joseph Cruikshank.
SOLD ALSO BY B. B. HOPKINS, & Co. Philadelphia;
D. MALLORY & Co. Boston; LYMAN, MALLORY &
Co. Portland; D. W. FARRAND & GREEN, Albany;
PH. H. NICKLIN, & Co. Baltimore; PATTERSON & HOP-
KINS, Pittsburgh.

....

1810.

(b)

Addition of Duodecimals.   169

2 What is the value of as many different dozens as may be
chosen out of 24 at 1d. per dozen?   ans 1126l. 6s. 4d.
3 How many different ways may a teacher select 50 sheep
out of a flock consisting of 100, so as not to make the same
choice twice?   ans. 1058913065445405275717249725256.

DUODECIMALS.

DUODECIMALS are fractions of a foot, or of an inch,
or parts of an inch, having 12 for their denominator.

The denominations are:

$$\begin{cases} 12 \text{ Fourths ''''} & \text{make 1 Third '''} \\ 12 \text{ Thirds '''} & 1 \text{ Second ''} \\ 12 \text{ Seconds ''} & 1 \text{ Inch } I. \\ 12 \text{ Inches} & 1 \text{ Foot } Ft. \end{cases}$$

ADDITION OF DUODECIMALS.

RULE.

Add as in compound addition, carrying one for each 12 to
the next denomination.

EXAMPLES.

Ft.	I.	''	'''	''''
14	4	6	5	6
85	6	6	6	9
56	10	5	7	9
43	11	6	4	3
87	11	10	8	5
48	5	2	10	11
336	5	1	7	4

	Ft.	I.	''	'''	''''
	28	4	3	7	10
	71	1	4	4	2
	67	11	5	1	5
	92	0	6	4	7
	43	3	8	11	10
	67	11	9	4	11

1 Five floors in a certain building contain each 1295f.
9i. 8'' how many feet in all?   answer 6479f. 0i. 4''.
2 Several boards measure as follow: viz. 27f. 3i. 25f. 11i.
23f. 10i. 20f. 9i. 20f. 6i. and 18f. 5i. what number of feet
do they contain?   answer 136f. 1i.

SUBTRACTION OF DUODECIMALS.

RULE.

Work as in compound subtraction, borrowing 12, when
necessary.

P

(c)

170   Multiplication of Duodecimals.

EXAMPLES.

	Ft.	I.	''	'''	''''
From	176	1	2	6	10
Take	97	10	1	7	11

	Ft.	I.	''	'''	''''
	3786	10	1	6	7
	987	8	11	6	9

Rem. 78 3 0 10 11
2 From a board measuring 41f. 7in. cut 19f. 10in. and
what is left?   answer 21f. 9in.

MULTIPLICATION OF DUODECIMALS.

CASE 1.

When the feet of the multiplier do no exceed 12;

RULE.

Set the multiplier in such order that the feet thereof may
stand under the lowest denomination of the multiplicand,
and in multiplying, carry one for every 12 from one denomination to another, and place the result of the lowest
denomination in the multiplicand under its multiplier.

Note 1. If there are no feet in the multiplier, supply their place with a
cipher.

2. Whether we begin with the highest or lowest denomination of the
multiplier, the several denominations of the products will be respectively synonymous with those of the multiplicand under which they
are placed.

EXAMPLES.

	Ft.	I.	''		Ft.	I.	''
Multiply	7	3	6		8	6	9 by 7 3 8
by 9f. 6in.	9	3					7 3 8
	3 10	6				5	8 6 0
	23	9				2	1 8 3
Product	27	1	6			59	11 3

CASE 2.

When the feet of the multiplier exceed 12;

RULE.

Use the component parts of the feet in the multiplier as in
compound multiplication, and take parts for the inches, &c.

	Ft.	I.	''	'''		Ft.	I.	''	'''
	62	6	7	9	0				

2 A mahogany board measures 28f. 10in. 6'' by 3f. 2in.
4'', what is its content?   answer 99f. 2in. 10'' 6''' 0''''

Figure 8-5   Title page and portions of the discussion of duodecimals from *The American Tutor's Assistant*, Philadelphia, 1810. The second edition was published in Philadelphia in 1791. It was written by John Todd, Zachariah Jess, William Waring, and Jeremiah Paul. No copy of the first edition is known.

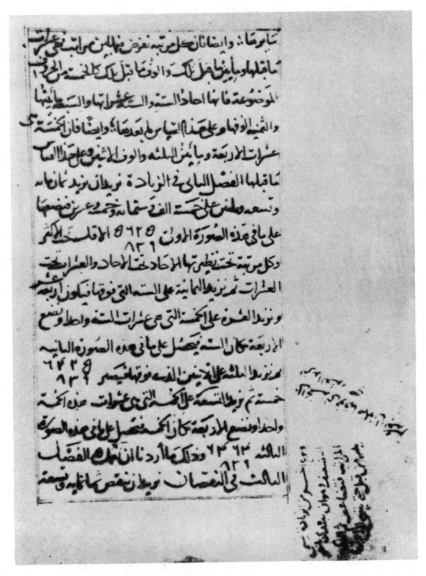

Figure 8-6 Folio 268b of the Manuscript of Kushyar Ibn Labban, *Principles of Hindu Reckoning,* published by the University of Wisconsin Press, 1965.

$$\frac{1}{3} = 0.333\ldots = 0.\overline{3}$$

$$\frac{1}{7} = 0.\overline{142857}$$

$$\frac{1}{8} = 0.125.$$

Of course, 0.5 can be written as 0.50000. . . or $0.5\overline{0}$, and $\frac{1}{8}$ as $0.125\overline{0}$. This shows that *every rational number can be represented as a repeating decimal.* Now the following two questions arise: Does every repeating decimal expression represent a rational number? Are there numbers represented by nonrepeating infinite decimals? The answer to both questions is "yes"; it is confirmed by modern theories of the real numbers in which the rational and the irrational numbers are combined in one unified system.

We shall use the repeating decimal $0.\overline{12}$ to develop a general algorithm for finding the $\frac{a}{b}$ form of the corresponding rational number. Let N represent this rational number. Then

$$N = 0.121212\ldots.$$

Multiplying both sides by 100 gives

$$100N = 12.121212\ldots.$$

Now we have

$$
\begin{array}{r}
100N = 12.121212\ldots \\
-\quad\ \ N = \ \ 0.121212\ldots \\
\hline
99N = 12.000000\ldots \\
= 12.
\end{array}
$$

Therefore,

$$N = \frac{12}{99} = \frac{4}{33}.$$

This result is easily verified by converting $\frac{4}{33}$ to a decimal.

Whenever the decimal expression repeats, this process works. One can always multiply the decimal expression by the power of 10 corresponding to the length of the period. The infinite succession of digits is then "moved over" to the left a number of places equal to the length of the period. Subtracting serves to eliminate the matched, repeating portions of the infinite "tails" on the two decimal expressions. The result has the form

$$aN = b$$

(a an integer, $a \neq 0$; b an integer or a terminating decimal), which gives, possibly after multiplication of a and b by a suitable power of 10,

$$N = \frac{d}{c}$$

(c and d integers, $c \neq 0$).

EXAMPLE 1

Let $N = 0.\overline{123}$. Then

$$1000N = 123.123123\ldots$$
$$-\quad N = \quad 0.123123\ldots$$
$$\overline{999N = 123}$$
$$N = \frac{123}{999} = \frac{41}{333}.$$

Check: Dividing 41 by 333 gives $0.\overline{123}$

EXAMPLE 2

Let $N = 0.7\overline{72}$. Then

$$100N = 77.272727\ldots$$
$$-\quad N = \quad 0.772727\ldots$$
$$\overline{99N = 76.5}$$
$$990N = 765$$
$$N = \frac{765}{990} = \frac{17}{22}.$$

Check: Dividing 17 by 22 gives $0.7\overline{72}$.

The results of Examples 1 and 2 suggest the theorem: *Every repeating decimal fraction represents a rational number.* There is one hidden flaw in the argument. We do not know if we may regard an infinite expression such as $0.121212\ldots$ as representing a single number to which we can apply the ordinary algorithms of arithmetic, such as the process for multiplying by 100 by "moving over" the digits two places. However, we will show another procedure for converting a repeating decimal to a common fraction, which avoids this difficulty.

The repeating decimal fraction

$$N = 0.12121212\ldots$$

can be written as an infinite series in the following way:

$$N = \frac{12}{100} + \frac{12}{10{,}000} + \frac{12}{1{,}000{,}000} + \frac{12}{100{,}000{,}000} + \cdots$$

$$= \frac{12}{100} + \frac{12}{100} \cdot \frac{1}{100} + \frac{12}{100} \cdot \left(\frac{1}{100}\right)^2 + \frac{12}{100} \cdot \left(\frac{1}{100}\right)^3 + \cdots.$$

Let us write the sum of the first n terms of this series, calling the result S_n. Thus,

$$S_n = \frac{12}{100} + \frac{12}{100} \cdot \frac{1}{100} + \frac{12}{100} \cdot \left(\frac{1}{100}\right)^2 + \ldots + \frac{12}{100} \cdot \left(\frac{1}{100}\right)^{n-1}$$

Then

$$\tfrac{1}{100}\, S_n = \frac{12}{100} \cdot \frac{1}{100} + \frac{12}{100} \cdot \left(\frac{1}{100}\right)^2 + \ldots + \frac{12}{100} \cdot \left(\frac{1}{100}\right)^{n-1}$$

$$+ \frac{12}{100} \cdot \left(\frac{1}{100}\right)^n.$$

It follows that

$$S_n - \tfrac{1}{100} S_n = \frac{12}{100} - \frac{12}{100} \cdot \left(\frac{1}{100}\right)^n$$

$$\tfrac{99}{100} S_n = \frac{12}{100} \left(1 - \frac{1}{100^n}\right)$$

$$S_n = \frac{12}{99} \left(1 - \frac{1}{100^n}\right).$$

The reader can see that as n gets larger $1/100^n$ gets smaller and approaches zero. Thus, S_n approaches $\frac{12}{99}$ as n gets larger. If we define N to be the number which is approached by S_n, then we see that

$$N = \frac{12}{99} = \frac{4}{33}.$$

In a similar way we can prove: *Every repeating decimal fraction represents a rational number.*

Thus, we have answered the first of the questions asked at the beginning of this section. The second question, Are there numbers represented by nonrepeating infinite decimals? has not been answered, but a complete answer is beyond the scope of this book. The steps that must be taken to complete the unified system of rational and irrational numbers are outlined below.

1. Every infinite decimal expression, whether repeating or not, represents a sequence that has a limit.
2. Such a limit can be regarded as *the definition of a real number.*

3. From (1) and (2) it follows that every infinite decimal expression represents a real number.

4. From (3) it follows that the expression 0.1010010001... represents a real number.

5. Every rational number can only be represented by a finite decimal expression or an infinite repeating decimal expression.

6. Since 0.1010010001. . . is neither finite nor infinite repeating, it follows from (5) that it does not represent a rational number.

7. From (4) and (6) it follows that there are real numbers which are not rational numbers.

The structure of the system of real numbers will be discussed in a more general setting in the next section.

Finally, then, we are back to the Pythagorean "scandal." The Pythagoreans showed that the diagonal and the side of a square are incommensurable (see Section 3-10). We now say that the ratio of the diagonal of a square to its side, that is, $\sqrt{2}$, is a real number which is not rational but which is represented by a nonrepeating infinite decimal expression, 1.41421.... Of course, if we cut off this decimal at a particular place, we have a representation of a rational number which is close to but not equal to $\sqrt{2}$.

EXERCISES 8-10

1 Whether the place-value representation of a fraction repeats or terminates depends on the base of the numeration system.

   (a) Give two examples of a place-value system in which $\frac{1}{3}$ can be represented by a terminating place-value expression. How would $\frac{1}{3}$ be represented in each of the systems?

   (b) Repeat (a), replacing the fraction $\frac{1}{3}$ by $\frac{2}{7}$.

   (c) A fraction in its lowest terms will be a "repeater" in a given base if and only if its denominator has a prime factor that is not a factor of the base (see Section 8-9). Complete the chart on page 253 by putting x's in the cells corresponding to a "repeater."

2 Find the decimal representation of

   (a) $\frac{1}{6}$ (b) $\frac{1}{7}$

   (c) $\frac{2}{7}$ (d) $\frac{3}{11}$

   (e) $\frac{6}{11}$ (f) $\frac{1}{12}$

   (g) $\frac{1}{13}$ (h) $\frac{20}{15}$

   (i) $\frac{21}{15}$ (j) $\frac{22}{15}$

3 (a) See Exercise 2. How can the answer to (c) be derived from the answer to (b)?

   (b) Find two other pairs which have the property that the answer to the second question can be derived from the answer to the first.

Base Fraction	2	5	7	10	12
$\frac{1}{2}$		X	X		
$\frac{1}{3}$	X	X	X	X	
$\frac{1}{4}$					
$\frac{1}{5}$					
$\frac{1}{6}$					
$\frac{1}{7}$					
$\frac{1}{8}$					
$\frac{1}{9}$					
$\frac{1}{10}$					
$\frac{1}{11}$					
$\frac{1}{12}$					
$\frac{1}{13}$					

(c) Discuss the differences in the results of (h), (i), and (j).

4 Find the common-fraction representation corresponding to the following repeating decimals:
    (a) 0.141414...   (b) $0.\overline{24}$
    (c) $0.4\overline{9}$ (d) $1.27\overline{6}$

5 A nine-decimal-place approximation to $\sqrt{2}$ is 1.414213562. Therefore, the best integer approximation to $\sqrt{2}$ is 1.
    (a) Find the best one-decimal-place, two-decimal-place, and three-decimal-place approximations to $\sqrt{2}$.
    (b) Check the precision of the approximations in (a) by squaring them and subtracting the results from 2.
    (c) Find the best seven-decimal-place approximation to $\sqrt{2}$.

6 The number π is an irrational number. An approximation to π is

$$3.14159265358979323846 2643.$$

Archimedes gave two approximations of π by proving:

$$3\tfrac{10}{71} < \pi < 3\tfrac{1}{7}$$

(See page 196). To how many decimal places does each of Archimedes' approximations agree with the correct expression?

8-11 MODERN THEORETICAL FOUNDATIONS
OF ARITHMETIC

We have seen that the idea of one-to-one correspondence of sets is implicit in primitive tallying (see page 223); we can see the idea of union of finite sets in the earliest appearance of addition of integers. However, the systematic development of a theory of sets and a clarification of the notion of cardinal number did not occur until the work of Georg Cantor in the latter part of the nineteenth century. In the seventeenth century the infinitesimal calculus had joined together studies of areas, volumes, curves, and centers of gravity with studies of velocity, acceleration, and force. The calculus was given a strong start by Leibniz and *Isaac Newton* (1643-1727). Throughout the eighteenth century mechanical algorithms and practical applications of the calculus developed rapidly, but the logical basis of the subject remained intuitive and incomplete. This led to some errors and to critical attacks by philosophers and mathematicians. Early in the nineteenth century *Augustin Cauchy* (1789-1857), *Bernard Riemann* (1826-1866), *Karl Weierstrass* (1815-1897), and other great mathematicians worked to build a solid logical structure for the theory of limits and the calculus. Cantor's work and that of his contemporary *Richard Dedekind* (1831-1916) filled a key position in these developments as well as in the clarification of the notion of real number and of the relationship between the real numbers and the points of a line.

One way to construct the real numbers *logically* from a few basic ideas is to start back with the natural numbers, the simplest and the oldest kind of number, and then, successively, to extend the system to include the integers, the rational numbers, and finally the real numbers.

It is interesting to note that the set of natural numbers was the last set of elements of the real-number system to be analyzed and reconstructed logically, although it was the first to grow up historically. *Giuseppe Peano* (1858-1932), an Italian logician and mathematician, formulated for his foundations of arithmetic a set of axioms in which he assumed the existence of a set of undefined elements, called *natural numbers*; an undefined element, *zero*; and the undefined notion of *successor*. All the properties of the natural numbers can be derived from these axioms.

The Peano axioms are:

1. Zero is a natural number.
2. If *a* is a natural number, then the successor of *a* is a natural number.
3. Zero is not the successor of a natural number.
4. If the successors of two natural numbers are equal, then the natural numbers themselves are equal.

5. If a set S of natural numbers contains zero and also the successor of every natural number in S, then every natural number is in S.

Today, some mathematicians prefer to replace "zero" by "one" in Peano's axioms. Then zero is not a natural number.

The first steps in developing Peano's theory were to name the natural numbers in sequence. According to the axioms, zero exists and has a successor. Call this successor "one." One is a natural number and therefore has a successor. Call this successor "two." And so forth. Thus, the ideas of order and of counting were made to grow *logically* out of Peano's axioms more than 5000 years after the first numbers were recorded.

Here, in Peano's theory, two lines of mathematical development, extending from antiquity, grew together. A Pythagorean notion (whole numbers are fundamental to everything) and an Aristotelian-Euclidean idea (the basic role of definitions and axioms) joined each other in the axiomatization of the system of the natural numbers in the nineteenth century.

The development of axioms for arithmetic and algebra is relatively recent. The Greeks developed a philosophical idea of an axiomatic system. In the fifth to third centuries B.C., Euclid and others axiomatized the geometry that had been developed from the intuitive and experimental ideas of earlier centuries. The concept of axiom changed in the forepart of the nineteenth century when non-Euclidean geometries were developed by changing Euclid's axioms.

Algebra, too, began to be axiomatized in the nineteenth century. The invention of new algebras paralleled the invention of new numbers and their arithmetics.

The study of structure in algebra is concerned with the investigation of the common properties of and the differences between algebraic systems. For example, the system of the rational numbers is an algebraic system which is the simplest example of an *ordered field*. The system of the real numbers and that of the complex numbers are also fields, but the complex number field is not ordered.

A detailed discussion of these matters is beyond the scope of this book. However, we have already mentioned many of the concepts associated with a field. Several of them have recently found their way into elementary and secondary school textbooks. We shall close, therefore, with two examples of a field, one ordered, the other not ordered.

A *field* is a set S of elements for which two binary operations are defined that satisfy the following conditions, the field axioms. (We use the letters a, b, c, ... to represent elements of the set. Since the most familiar fields involve real numbers and the ordinary operations of addition and multiplication, we shall use the symbols $\oplus$ and $\odot$ for these operations.)

Addition

Multiplication

1. *S* is *closed* under addition. If *a* and *b* belong to *S*, then $a \oplus b$ belongs to *S*.

2. Addition is *associative.* $(a \oplus b) \oplus c = a \oplus (b \oplus c)$.

3. Addition is *commutative.* $a \oplus b = b \oplus a$.

4. There is an *additive identity element* in *S*. We indicate this element by 0. $a \oplus 0 = a$.

5. For each element *a* of *S* there is an *additive inverse,* represented by $-a$. $a \oplus -a = 0$.

1. *S* is *closed* under multiplication. If *a* and *b* belong to *S*, then $a \odot b$ belongs to *S*.

2. Multiplication is *associative.* $(a \odot b) \odot c = a \odot (b \odot c)$.

3. Multiplication is *commutative.* $a \odot b = b \odot a$.

4. There is a *multiplicative identity element* in *S*. We indicate this element by 1. $a \odot 1 = a$.

5. For each element *a* of *S*, except 0, there is a *multiplicative inverse,* represented by a^{-1}. $a \odot a^{-1} = 1 \qquad (a \neq 0)$.

6. Multiplication *distributes* over addition. $a \odot (b \oplus c) = (a \odot b) \oplus (a \odot c)$.

7. $1 \neq 0$.

We have mentioned the fact that the system of the rational numbers is an example of a field. It is fairly obvious from experience that the rational numbers satisfy the above 12 conditions.

There are many other fields, such as the finite fields, which have been studied recently. For example, take as a set of elements the set of numbers {0, 1, 2, 3, 4}, and let the following tables define the two operations $\oplus$ and $\odot$:

$\oplus$	0	1	2	3	4
0	0	1	2	3	4
1	1	2	3	4	0
2	2	3	4	0	1
3	3	4	0	1	2
4	4	0	1	2	3

$\odot$	0	1	2	3	4
0	0	0	0	0	0
1	0	1	2	3	4
2	0	2	4	1	3
3	0	3	1	4	2
4	0	4	3	2	1

The set, together with the operations defined by the tables, is called a *finite field modulo 5*. By checking numerical examples, verify that this system satisfies the field axioms. For instance, the additive inverse of 2 is found by reading across the first table opposite 2 until a zero (the additive identity) is found. The fact that $2 + 3 = 0$ shows that 3 is the additive inverse of 2. The distributive axiom is illustrated as follows:

$$2 \odot (3 \oplus 4) = 2 \odot 2 = 4$$
$$(2 \odot 3) \oplus (2 \odot 4) = 1 \oplus 3 = 4.$$

Therefore,

$$2 \odot (3 \oplus 4) = (2 \odot 3) \oplus (2 \odot 4).$$

Figure 8-7

In elementary mathematics texts a finite field of this type is called a *modular arithmetic.* It is also called a "clock arithmetic" because the addition table can be derived by adding "hours" on the five-hour clock shown in Figure 8-7. If we start at 2 on this clock and go forward three spaces we arrive at 0. That is, in our addition table, $2 \oplus 3 = 0$. The multiplication table can be derived in several ways. One way uses repeated addition on the clock. Another uses ordinary multiplication followed by subtraction of 5's. Thus,

$$2 \cdot 3 = 6$$

hence,

$$2 \odot 3 = 6 - 5 = 1;$$

$$3 \cdot 4 = 12$$

hence,

$$3 \odot 4 = 12 - 2 \cdot 5 = 2.$$

We have seen that this modular arithmetic and the system of rational numbers are fields. However, the rational numbers have an additional set of properties, the *order properties*, not enjoyed by finite fields. We merely illustrate this difference. If the symbol $>$ means "is greater than," then $5 > 3$ is read "5 is greater than 3."

"Greater than" has the following properties:

1. Of two numbers, a and b, exactly one of

   $$a > b, \quad b > a, \quad a = b$$

   is true.

2. If

   $$a > b \quad \text{and} \quad b > c,$$

   then

   $$a > c.$$

3. If

   $$a > b,$$

   then

   $$a + c > b + c.$$

We can easily verify that the rational numbers have the order properties.

Now note what would happen in modulo 5 arithmetic if we would use a relation "is greater than" with the properties (1), (2), and (3). First assume that $1 > 0$. Then we have

$$1 + 1 > 0 + 1 \qquad \text{(property (3))}.$$

Hence, $2 > 1,$

and since $1 > 0,$

we have $2 > 0 \qquad \text{(property (2))}.$

Similarly, $3 > 0$

                     $4 > 0$

                     $0 > 0,$

which is contrary to property (1). In the same way we can show that the assumption $0 > 1$ leads to $0 > 0$. Since the assumptions $1 > 0$ and $0 > 1$ both lead to a contradiction, it then follows from property (1) that $1 = 0$, contrary to Axiom 7 (page 256).

This example illustrates the reasons a finite field cannot be an ordered field.

The perception that the properties of the rational numbers can be derived logically from the axioms for an ordered field developed in the latter part of the nineteenth century. Dedekind described a special case of a field in 1871, and *Heinrich Weber* (1842–1913) formulated abstract approaches in 1882 and 1893. The American mathematician *E. H. Moore* also presented an abstract formulation in the latter year. Finite fields, especially in certain advanced aspects, are largely a twentieth-century development. These were important developments, not only because they provided better understanding and simpler and clearer structures for many old systems, but also because they suggested and stimulated the development of many new arithmetic and algebraic systems, some of them being radically different from classical mathematics. These differences (some new algebras are noncommutative or nonassociative) are not only interesting, but they have been important in stimulating further developments and have even found some unforseen applications. For example, noncommutative vector and quaternion algebras, although not fields, have played important roles in mechanics and other parts of physics and engineering.

This section has been designed to illustrate how elementary mathematics has come such a long way over the centuries. Our sketch points out that many of the elements of modern mathematical structures grew up quite naturally in antiquity. It has been an achievement of modern mathematics to "put it all together," to display the essential elements as parts of a comprehensive structure, and to go on to invent entirely new systems.

EXERCISES 8-11

1 Show that the set consisting of the elements *a*, *b*, and *c*, on which the operations ⊕ and ⊙ are defined by the following tables, constitutes a field:

⊕	a	b	c		⊙	a	b	c
a	a	b	c		a	a	a	a
b	b	c	a		b	a	b	c
c	c	a	b		c	a	c	b

2 Use the tables of the finite field modulo 5, given in this section, to solve the following equations.

(a) $2 \oplus x = 4$ (b) $2 \oplus x = 0$
(c) $2 \oplus x = 1$ (d) $2 \odot x = 4$
(e) $2 \odot x = 0$ (f) $2 \odot x = 1$

3 Replace *a*, *b*, and *c* in the tables in Exercise 1 by 0, 1, and 2. Use the resulting tables to solve the following equations.

(a) $2 \oplus x = 0$ (b) $2 \oplus x = 1$
(c) $2 \oplus x = 2$ (d) $2 \odot x = 0$
(e) $2 \odot x = 1$ (f) $2 \odot x = 2$

4 Solve the equations in Exercise 3 in the modulo 5 system.

5 Make ⊕ and ⊙ tables for a four-hour clock (using 0, 1, 2, 3) and show that this system is not a field.

8-12 MODERN NUMERATION

For our final section we return to the topic of our first section, numeration. Today most of the world's business and science use the Hindu–Arabic numerals even when the accompanying discussion is written in an alphabet different from our own. However, the Arabic and Indian countries use a different form of these numerals. Table 8-1 shows a comparison of Ibn Labban's numerals with those in use today in Arabic and Indian regions. Figure 8-8 shows some of these numerals as they appear on modern stamps.

Table 8-1

	0	1	2	3	4	5	6	7	8	9
Ibn Labban (971-1029)	•	١	٢	٣	٣	٩	٦	٧	٨	٩
Arabic (today)	٠	١	٢	٣	٤	٥	٦	٧	٨	٩
Hindi (today)	٥	१	२	३	४	५	६	७	८	९

Figure 8-8 Egyptian stamps showing modern Arabic numerals.

Big numbers, millions, billions, trillions, and so on, have always had a fascination for both children and adults. In *The Sand Reckoner,* Archimedes showed how the number of grains of sand in a universe full of sand could be computed. He was challenged by the problem of whether such a large number could be written and used in computation. We must keep in mind that the Greek numeration system lacked the place-value notion of our modern system.

The ancient Hindus, too, found large numbers fascinating. They gave separate names to each decimal place. The following excerpt from Edwin Arnold's *The Light of Asia* is part of a legendary account of the education of the young Buddha at the hands of his teacher Viswamitra. It indicates the importance of number names and large numbers in early Hindu arithmetic.

> And Viswamitra said, "It is enough,
> Let us to numbers.
>   After me repeat
> Your numeration till we reach the Lakh,
> One, two, three, four, to ten, and then by tens
> To hundreds, thousands." After him the child
> Named digits, decads, centuries; nor paused,
> The round lakh reached, but softly murmured on,
> "Then comes the koti, nahut, ninnahut,
> Khamba, viskhamba, abab, attata,
> To kumuds, gundhikas, and utpalas,
> By pundarikas unto padumas
> Which last is how you count the utmost grains

Of Hastagiri ground to finest dust;
But beyond that a numeration is,
The Katha, used to count the stars of night;
The Koti-Katha, for the ocean drops;
Ingga, the calculus of circulars;
Sarvanikchepa, by the which you deal
With all the sands of Gunga, till we come
to Antah-Kalpas, where the unit is
The sands of ten crore Gungas. If one seeks
More comprehensive scale, th' arithmic mounts
By the Asankya, which is the tale
Of all the drops that in ten thousand years
Would fall on all the worlds by daily rain;
Thence unto Maha Kalpas, by the which
The gods compute their future and their past."
. . . .
. . . . "and, Master! If it please,
I shall recite how many sun-motes lie
From end to end within a yojana."
Thereat with instant skill, the little Prince
Pronounced the total of the atoms true.
But Viswamitra heard it on his face
Prostrate before the boy; "For thou," he cried,
"Art Teacher of thy teachers—Thou, not I,
Art Guru."

Some years ago newspapers and magazines carried articles on the new words *googol* (1, followed by 100 zeros) and *googolplex* (1, followed by a googol of zeros.) The words grew out of a discussion between Edward Kasner of Columbia University and some kindergarten children concerning the numerals needed to describe the number of raindrops falling on New York City in a spring rain. They saw that this was a very large number, but still a finite number, which could be named and written. The words googol and googolplex were invented by Kasner's nine-year-old nephew.

The numbers mentioned above can be written conveniently by the use of *standard* (or scientific) *notation*. In this notation every number is written as a number between 1 and 10, multiplied by a power of 10. Thus,

$$45 = 4.5 \times 10^1$$
$$452 = 4.52 \times 10^2$$
$$4,526,392,000 = 4.526392 \times 10^9.$$

We now use this idea to write (note the difference between American and English usage):

$$\text{an American billion} = 1,000,000,000 = 10^9$$
$$\text{an English billion} = 1,000,000,000,000 = 10^{12}$$
$$\text{an American trillion} = 10^{12}$$
$$\text{an English trillion} = 10^{18}$$
$$\text{a googol} = 10^{100}$$
$$\text{a googolplex} = 10^{10^{100}}$$

Scientists also need to express very small numbers. Then they use negative exponents. Thus,

$$0.45 = 4.5 \cdot \frac{1}{10} = 4.5 \times 10^{-1}$$

$$0.045 = 4.5 \cdot \frac{1}{10^2} = 4.5 \times 10^{-2}$$

$$0.000000563 = 5.63 \cdot 10^{-7}$$

EXERCISES 8-12

1 Write four trillion
   (a) using the American definition and writing all the zeros,
   (b) using the English definition and writing all the zeros,
   (c) using the American definition and standard notation,
   (d) using the English definition and standard notation.

2 (a) The average distance from the sun to the earth is 9.3×10^7 miles. Write the complete numeral for this number and also its name in words.
   (b) The distance from the earth to one of the farthest stars is 8,225,000,000, 000,000,000 miles. Write this distance in standard notation.

3 (a) The diameter of a nitrogen molecule is 1.8×10^{-8} centimeter. Write the complete numeral for this number and also its name in words.
   (b) Write 0.00000000027 in standard notation.

4 Although it is hard to visualize large numbers, people try various devices to describe them. The January 1952 *Reader's Digest* quotes James L. Holton as writing: "A million dollars in new $1000 bills would make a pile 8 inches high. But if we tried to pile up a billion dollars, we'd find it stretched up in the sky 110 feet higher than the Washington Monument." Measure the thickness of a bill using a micrometer and check out Holton's statement.

5 What is the largest number that one can say in English? How did the Romans write large numbers? Answers to these questions may be found in reference 46. A chart or poster of the history of big numbers, their names, symbols, and uses, would make an interesting display.

REFERENCES

For more about number symbols, see

[43] Cajori, Florian, *A History of Mathematical Notations.* Chicago: Open Court Publishing Company, 1928.
[44] Menninger, Karl, *Number Words and Number Symbols.* Cambridge, Mass. The MIT Press, 1969.

For a discussion of tangible arithmetic, see

[45] *The Mathematics Teacher,* Vol. 47 (1954), pp. 482–487, 535–542; Vol. 48 (1955), pp. 91–95, 153–157, 250.

For a discussion of large numbers and their names, see

[46] *The Mathematics Teacher,* Vol. 45 (1952), pp. 528–530; Vol. 46 (1953), pp. 265–269; Vol. 47 (1954), pp. 194–195 (note the correction on p. 345).

HINTS AND ANSWERS
TO SELECTED EXERCISES

*Where proofs or methods of thinking are required, we have not attempted
to supply all the steps or alternate solutions. We encourage our readers
to be different and inventive. Wherever errors are found we would be
pleased to be told.*

EXERCISES 1-3

2 (a) ∩∩ III.   (b) ℮℮ III ℮℮ IIII·   (d) 𝈇 ⚎⚎ ℮℮℮ ∩∩ II ∩∩ III·

3 (a) 20,507.   (b) 2,000,000.   (c) 100,000.

4 7. 10.

5 𝈇 ℮ ∩ II III· None. 45.

6 ⌒ 𝈇𝈇 ℮℮ ℮℮ ∩∩∩. Change each symbol to one of the next higher
denomination.

EXERCISES 1-4

1 (a) ∩∩ III
       ∩∩ III
       ∩∩ III
       ─────────
       ∩∩∩ IIII
       ∩∩∩ IIIII

3 (a)   1   17
        \   2   34
        \   4   68
              8   136
        \   16   272   sum 374.

EXERCISES 1-5

1 (c)   1   692
        \   2   1384
              4   2768
        \   8   5536
        \   16   11072
        \   32   22144   sum 40,136.

2 (c)   58 X   692
         29 X   1384   /
         14 X   2768
          7 X   5536   /
          3 X   11072   /
          1 X   22144   /
         Answer: 40136

3   Yes. If the larger number is used for the multiplier, the problem
    requires 9 doublings; if the smaller number is used, only 5 doublings
    are required.

5 (c)   1   11
        \   2   22
        \   4   44
              8   88
        \   16   176   sum 242.
        Hence, $242 \div 11 = 2 + 4 + 16 = 22.$

6   (a) $15 = 2^3 + 2^2 + 2^1 + 2^0.$ (c) $22 = 2^4 + 2^2 + 2^1.$
    (e) $16 = 2^4.$ (g) $968 = 2^9 + 2^8 + 2^7 + 2^6 + 2^3.$

EXERCISES 1-6

1 (a) \   1   20
             $\overline{2}$ 10
        \ $\overline{4}$ 5
        \ $\overline{20}$ 1   sum 26.
        Hence, $26 \div 20 = 1 + \overline{4} + \overline{20}.$

(b) \\ 1   6
        2   12
        4   24
     \\ 8   48
     \\ $\bar{6}$ 1   sum 55.

Hence, $55 \div 6 = 1 + 8 + \bar{6} = 9 + \bar{6}$.

(c) \\ 1   21
     \\ 2   42
         $\bar{\bar{3}}$ 14
     \\ $\bar{3}$ 7
     \\ $\overline{21}$ 1   sum 71.

Hence, $71 \div 21 = 1 + 2 + \bar{3} + \overline{21}$
$= 3 + \bar{3} + \overline{21}$.

(d) $1 + \bar{3} + \overline{18}$.   (e) $\bar{2} + \bar{4} + \overline{68}$.   (f) $\bar{3} + \overline{36}$.

2 (a)   1   4
     \\ $\bar{2}$ 2
     \\ $\bar{4}$ 1   sum 3.

Hence, $3 \div 4 = \bar{2} + \bar{4}$.

(b)   1   8
     \\ $\bar{2}$ 4
         $\bar{4}$ 2
     \\ $\bar{8}$ 1   sum 5.

Hence, $5 \div 8 = \bar{2} + \bar{8}$.

(c)   1   24
         $\bar{\bar{3}}$ 16
     \\ $\bar{3}$ 8
     \\ $\bar{6}$ 4
     \\ $\overline{12}$ 2   sum 14.

Hence, $14 \div 24 = \bar{3} + \bar{6} + \overline{12}$.

(d) $1 + \overline{16} + \overline{32}$.   (e) $\bar{3} + \bar{6}$.   (f) $1 + \bar{3} + \overline{12}$.   (g) $\bar{2} + \bar{8} + \overline{16}$.

(h) $2 + \bar{\bar{3}} + \bar{6}$.

3 (a) $\dfrac{1}{3} + \dfrac{1}{36}$.   (c) $\dfrac{1}{4} + \dfrac{1}{60}$.

4 (a) $\dfrac{1}{4} + \dfrac{1}{9}$.   (c) $\dfrac{1}{5} + \dfrac{1}{15}$.

5 (c) If m is even then $\dfrac{2}{m} = \dfrac{2}{2n} = \dfrac{1}{n}$, n an integer, and $\dfrac{1}{n}$ is already a unit fraction.

6 (c) Let $\dfrac{a}{b} = \dfrac{1}{a_1} + \dfrac{1}{a_2} + \ldots + \dfrac{1}{a_r}$, $a_1 < a_2 < \ldots < a_r$.

Apply part a to $\dfrac{1}{a_r}$.

7 Let $n = am + b$, $1 \leqq b < m$. Then

$$\frac{m}{n} = \frac{1}{a+1} + \left(\frac{m}{n} - \frac{1}{a+1}\right) = \frac{1}{a+1} + \left(\frac{m}{am+b} - \frac{1}{a+1}\right)$$

$$= \frac{1}{a+1} + \frac{m-b}{(am+b)(a+1)}, \text{ with } b \geqq 1. \text{ In the last expression, the}$$

second fraction has a smaller numerator than the numerator we began with. Therefore, the procedure terminates after a finite number of steps.

8 The proof of Exercise 7 shows that the representation is produced by a well-defined algorithm.

EXERCISES 1-7

1 (a) $\overline{12}$ $\overline{6}$ $\overline{3}$

    1   2   4   sum 7, remainder 5. Calculate

with 12 until you find 5.

$$\begin{array}{cc}
1 & 12 \\
\overline{3} & 8 \\
\backslash \quad \overline{3} & 4 \\
\backslash \quad \overline{12} & 1 \qquad \text{sum 5.}
\end{array}$$

Hence, $\dfrac{5}{12} = \overline{3} + \overline{12}$ and $(\overline{12} + \overline{6} + \overline{3}) + (\overline{3} + \overline{12}) = 1$.

(b) $\overline{6} + \overline{24} + \overline{120}$.   or $\overline{5} + \overline{60}$.   (c) $\overline{2} + \overline{28}$.   (d) $\overline{3} + \overline{18}$.

(e) $\overline{12}$.   (f) $\overline{2} + \overline{4}$.   (g) $\overline{3} + \overline{12} + \overline{24}$.

(h)

$$\begin{array}{cccc}
1 & 60 \\
\backslash \quad 2 & 30 \\
4 & 15 \\
\backslash \quad 8 & 7 & \overline{2} \\
\overline{16} & 3 & \overline{2} & \overline{4} \\
\overline{32} & 1 & \overline{2} & \overline{4} & \overline{8} \\
\backslash \quad \overline{64} & & \overline{2} & \overline{4} & \overline{8} & \overline{16}
\end{array}$$

Complete $\overline{4} + \overline{8} + \overline{16}$ to 1. The computation is shown at the right.

$\overline{4} + \overline{8} + \overline{16}$

4   2   1   sum 7, remainder 9. Calculate with 16 until you find 9.

$$\begin{array}{cc}
1 & 16 \\
\backslash \quad 2 & 8 \\
\backslash \quad \overline{16} & 1 \qquad \text{sum 9.}
\end{array}$$

Hence, $9 \div 16 = \overline{2} + \overline{16}$.

$$\begin{array}{ll} & \overline{60} & 1 \\ \backslash & \overline{120} & \overline{2} \\ \backslash & \overline{960} & \overline{16} \end{array}$$

Hence, the *skm* is $\overline{2} + \overline{8} + \overline{64} + \overline{120} + \overline{960}$.

2 (a)
$$\begin{array}{llll} & 1 & 23 & \\ \backslash & \overline{2} & 11 & \overline{2} \\ & \overline{23} & 1 & \\ \backslash & \overline{46} & \overline{2} & \end{array}$$

  sum $11 + \overline{2}$. We need $\overline{2}$ more.

  sum 12.

Hence, $12 \div 23 = \overline{2} + \overline{46}$.

(b) $\overline{3} + \overline{6} + \overline{78}$.   (c) $\overline{2} + \overline{4} + \overline{38} + \overline{76}$.   (d) $4 + \overline{3} + \overline{21}$.
(e) $\overline{6} + \overline{390}$.   (f) $\overline{3} + \overline{24} + \overline{92} + \overline{184}$.

3 Suppose $\dfrac{5}{17} = \dfrac{1}{2^{n_1}} + \dfrac{1}{2^{n_2}} + \dots + \dfrac{1}{2^{n_k}}$. Then

$$\frac{5}{17} = \frac{\text{(sum of powers of 2)} + 1}{2^{n_k}} = \frac{\text{odd number}}{\text{even number}}.$$

Then $5 \times$ (even number) $= 17 \times$ (odd number), and hence
(even number) $=$ (odd number), which is impossible.

EXERCISES 1-8

1 (a)
$$\begin{array}{lllll} & 1 & 1 & \overline{5} & \\ & 2 & 2 & \overline{3} & \overline{15} \\ \backslash & 4 & 4 & \overline{3} & \overline{10} & \overline{30} \end{array}$$
    (from $2 \div n$ table)
    (from $2 \div n$ table)

Hence, the quotient is 4.

(b) 8.

EXERCISES 1-9

1 (a) ꝏꝫ·ꝪꞮ (b) IꞮꞮꞮ⁄ (c) Iꟷꟷ (d) ꞬꞮꞮꞮ⁄

2 $\dfrac{1}{21} + \dfrac{1}{42} = \dfrac{1}{14}$.

3 $\dfrac{1}{24} + \dfrac{1}{48} = \dfrac{1}{16}$.

EXERCISES 1-10

1 (a) Assume 2. Then $2 + \dfrac{1}{2} \cdot 2 = 3$. $3 \neq 16$; hence we calculate
$16 \div 3 = 5\tfrac{1}{3}$. $5\tfrac{1}{3} \cdot 2 = 10\tfrac{2}{3}$. (b) 9.

2 (a) $1\frac{2}{3}$, $10\frac{5}{6}$, 20, $29\frac{1}{6}$, $38\frac{1}{3}$. (b) (1) Find the sum of the first 2 shares by using false position; answer: $12\frac{1}{2}$. Then the sum of the last 3 shares is $87\frac{1}{2}$. (2) Twice the sum of the last 3 minus thrice the sum of the first 2 is $137\frac{1}{2}$, which is also twice 9 times the common difference minus thrice the difference. (3) Find the common difference by using false position; answer: $9\frac{1}{6}$. (4) Twice the first share plus one difference is $12\frac{1}{2}$. Hence, the first share is $1\frac{2}{3}$.

EXERCISES 1-11

1 (a) 55,986.

2 (a) $\frac{63}{64}$.

4 (a) $113\frac{7}{9} \approx 113.78$. (b) 113.04. (c) 0.7%.

EXERCISES 2-2

1 (a) 2; $1,0,1 = 3601$; $2,0,0 = 7200$.

2 (a) 1.375. (b) $12,23 = 743$.

3 (a) 1,22;30. (b) Move the sexagesimal point one position to the right.

4 (a) 〔cuneiform〕 (c) 〔cuneiform〕 (e) 〔cuneiform〕

5 (a) 〔cuneiform〕 (c) 〔cuneiform〕 (e) 〔cuneiform〕

EXERCISES 2-3

1 $1 \div 12 = 0;5$
$1 \div 15 = 0;4$
$1 \div 16 = 0;3,45$
etc.
$1 \div 54 = 0;1,6,40$
$1 \div 60 = 0;1$

2 $9 \times 0;6,40 = 1$
$10 \times 0;6,40 = 1;6,40$
$11 \times 0;6,40 = 1;13,20$
etc.
$18 \times 0;6,40 = 2$.
(c) $25 \div 9 = 20 \times 0;6,40 + 5 \times 0;6,40$
$= 2;13,20 + 0;33,20 = 2;46,40$.

3 (a) Make up multiplication table of $\frac{1}{5} = 0;12$ until $18 \times 0;12 = 3;36$.

(c) Make up multiplication table of $\frac{1}{10} = 0;6$ until $17 \times 0;6 = 1;42$.

EXERCISES 2-4

1 (a) 3, $3\frac{1}{6}$, $3\frac{37}{228}$. (b) 3, $2\frac{2}{3}$, $2\frac{31}{48}$. (c) 4, $3\frac{7}{8}$, $3\frac{433}{496}$.

(d) 5, $5\frac{1}{5}$, $5\frac{51}{260}$.

2 First approximation 40.

Second approximation $\dfrac{40 + (40^2 + 10^2)/40}{2} = 40 + \dfrac{100}{2 \cdot 40}$.

EXERCISES 2-5

1 $(15 + a)(15 - a) = 104$; $a = 11$. Answer: $4, 26$.

3 (b) They did not know negative numbers. (c) $-1, -29$.

5 $(3x)^2 + 4(3x) = 12$, etc.; $x = \dfrac{2}{3}$.

EXERCISES 2-9

1 $a = 24$, $b = 10$, $c = 26$; $24^2 + 10^2 = 26^2$.

2 $2,0$.

3 (a) $1,12$. (b) $1,0$.

4 (a) 30. (b) $31\frac{1}{4}$. (c) 31.416.

(d) Error: 1.416. Percent: $\dfrac{1.416}{0.31416} \approx 4\frac{1}{2}$.

5 $4\frac{1}{2}$.

6 (d) Error: 3.54. Percent: $4\frac{1}{2}$.

7 (d) Error: 14.16. Percent: $4\frac{1}{2}$.

8 $R = 10$. Let M be the center of the circle. Apply the theorem of Pythagoras to $\triangle MBC$, in which $MB = 8$ and $MC = 10$.

Answer: $CD = 12$.

9 $\left.\begin{array}{l} \frac{1}{2}x(y_2 - y_1) + 15\,y_2 = 420 \\ y_1 - y_2 = 20 \end{array}\right\} \Rightarrow \begin{array}{l} -10x + 15y_2 = 420 \\ y_1 - y_2 = 20 \end{array}$

$x = \dfrac{15y_2 - 420}{10}$, $y_1 = y_2 + 20$

Substitute for x and y_1 in $30\,y_1 = x\,(y_1 + y_2)$. Then solve for y_2.
Answer: $x = 18$, $y_1 = 60$, $y_2 = 40$.

EXERCISES 3-2

1   (a) $\kappa\gamma$.   (b) $\rho\zeta$.   (c) $\sigma\kappa\zeta$.   (d) $,\eta o\nu\varsigma$.   (e) $\overset{o\varsigma}{M},\theta\tau\epsilon$.

(f) $\gamma'\,\epsilon''\,\epsilon''$ or $\overset{\epsilon}{\gamma}$.   (g) $\iota\theta'\kappa\alpha''\,\kappa\alpha''$ or $\overset{\kappa\alpha}{\iota\theta}$.

(h) $\tau\kappa\eta'\,\varphi\zeta''\,\varphi\zeta''$ or $\overset{\varphi\zeta}{\tau\kappa\eta}$.

2   (a) 35.   (c) 566.   (e) 856,083.   (g) $\dfrac{30}{45}$.   (i) $\dfrac{35}{41}$.

3   (a) $\varphi\nu\zeta$.   (b) $\overset{\epsilon}{M},\zeta\sigma\alpha$.

4   $\theta\,\omega\xi\delta$.

5   (a) △△||| (c) HH△△Γ||

EXERCISES 3-3

1   By proving $\triangle SWM \cong \triangle QPM$.

2   Theorems 2 and 3.

EXERCISES 3-5

1   (a) 9; $8\frac{5}{9}$.   (c) -3; -3.

2   (a) Yes.   (b) No.

3   (a) $23\frac{1}{6}$.

4   (a) $5\frac{1}{7}$.   (b) $10\frac{2}{7}$.   (c) $6\frac{174}{391}$.

EXERCISES 3-6

2   $T_n = 1 + 2 + 3 + \ldots + n = \frac{1}{2}n\,(n+1)$.

3   Separate the set of dots representing P_n into 4 subsets: one of n dots and three of T_{n-1} dots each. $P_n = n + 3\,T_{n-1}$
$= n + 3 \cdot \frac{1}{2}\,(n-1)\,n = \frac{1}{2}n\,(3n-1)$.

4 (b)

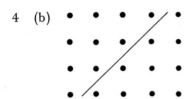

5 (a)

(b) $T_{n-1} + T_n = \frac{1}{2}(n-1)n + \frac{1}{2}n(n+1) = n^2 = S_n.$

6 (c) The procedure does not produce the triple $(6, 8, 10)$.

7 (b) $1 + 3 + 5 + \ldots + (1 + (n-1)\cdot 2) = \frac{1}{2}n(1 + (2n-1)) = n^2.$

8 For $n = 3$: $2^{n-1}(2^n-1) = 28$. Proper divisors $1, 2, 4, 7, 14$.
   Their sum is 28. For $n = 6$: the number is 2016. The sum of the
   largest 4 proper divisors is greater than 2016.

9 (a) $n = 7$ gives $2^6(2^7-1) = 2^{13} - 2^6 \approx 2^{13}$; *etc.*   Answer: 4.
   (c) 77.

10 (a) Third theorem: $2n + 1$ divides $2m$; $2m = k(2n+1)$;
        k is even $= 2l$; $2m = 2l(2n+1)$; $m = l(2n+1)$; $2n+1$ divides m.

EXERCISES 3-10

1 (a) n is even or odd. n odd $\Rightarrow n^2$ odd, since $(2m + 1)^2 =$
        $2(2m^2 + 2m) + 1$. Hence, n^2 even $\Rightarrow n$ even.
   (b) Follows from part (a).

3 Suppose $5 + \sqrt{2} = r$, r rational. Then $\sqrt{2} = r - 5$, a rational number.

5 Suppose $\sqrt{3} = \frac{p}{q}$, p and q integers with no common factor. Then
   $p^2 = 3q^2$, which implies p^2 is divisible by 3. p must be of one of
   the forms $3k$, $3k + 1$, or $3k + 2$. p cannot be of the form $3k + 1$
   because then $p^2 = (3k + 1)^2 = 3(3k^2 + 2k) + 1$ which is not

divisible by 3. Show similarly that p is not of the form $3k + 2$. Then p is of the form $3k$. From this show 3 also must divide q which contradicts the assumption that p and q have no common factors.

EXERCISES 4-2

1 $\dfrac{AC^2}{AB^2} = \dfrac{AC^2}{AC^2 + BC^2} = \dfrac{AC^2}{2AC^2} = \dfrac{1}{2}.$

2 (a)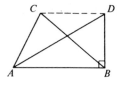

   $CD \parallel AB$
   Area $\triangle ABD$ = area $\triangle ABC$

   (b)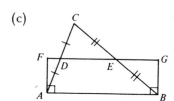

   Area rectangle $ABFD$ = area $\triangle ABC$

   (c)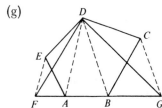

   Area rectangle $ABGF$ = area $\triangle ABC$

   (d) $\triangle DCG \sim \triangle GCH.$ $\therefore \dfrac{DC}{CG} = \dfrac{CG}{CH}$, or $\dfrac{l}{s} = \dfrac{s}{w}.$

   (f) Construct a rectangle equal in area to the given trapezoid. Then apply part (d).

   (g) Construct $\triangle FGD$ equal in area to pentagon $ABCDE$ ($EF \parallel DA$, $CG \parallel DB$). Then apply parts (b) and (d).

   (h) Apply the theorem of Pythagoras.

3   The side of the square has length 18.

4   (a) 56.   (c) $34\frac{1}{2}$.

EXERCISES 4-4

1   (1) To prove formula (4), drop the perpendicular $\overline{AF}$ from A onto the extension of $\overline{CD}$. In $\triangle AFC$, $AC^2 = AF^2 + FC^2 = (AD^2 - DF^2) + (DF + DC)^2 = AD^2 + DC^2 + 2DF \cdot DC > AD^2 + DC^2$.
(2) Since $AB^2 < AC^2 + CB^2$, $\angle ACB$ in $\triangle ABC$ cannot be a right angle; it cannot be an obtuse angle either, which can be proved in a way similar to the way we proved that $AC^2 > AD^2 + DC^2$.
$\therefore$ $\angle ACB$ is an acute angle.
(3) Complete circle ACB. Then m (minor arc AB) $= 2m \angle ACB < 180°$. $\therefore$ The major arc AB is greater than a semicircle.

2   (c) If $\overline{CF} \perp \overline{AB}$ (F on $\overline{AB}$), $\triangle FBC$ can be constructed since $FB = \frac{1}{2}(AB - CD)$.
(e) Let M be the center of the circle through A, B, and C. Construct a triangle ABN (N on the other side of $\overline{AB}$ from C), such that $\triangle ABN \sim \triangle ADM$.

3   (a) We first derive the area of a segment corresponding to a central angle of measure α (see figure).

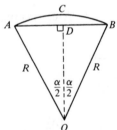

$$\frac{\text{area sector } AOBC}{\text{area} \odot O} = \frac{\alpha}{360}.$$

$\therefore$ Area sector $AOBC = \dfrac{\alpha}{360} \cdot \pi R^2$.

Area $\triangle AOB = \frac{1}{2}AB \cdot OD = R \sin \frac{\alpha}{2} \cdot R \cos \frac{\alpha}{2} = R^2 \sin \frac{\alpha}{2} \cos \frac{\alpha}{2}$.

$\therefore$ Area segment $ABC = \left(\dfrac{\alpha \pi}{360} - \sin \dfrac{\alpha}{2} \cos \dfrac{\alpha}{2} \right) R^2$.

Now let $A_1 O_1 B_1 C_1$ be a sector in a circle of radius R_1, also corresponding to a central angle of measure α. Then

$$\frac{\text{area segment } ABC}{\text{area segment } A_1 B_1 C_1} = \frac{\left(\dfrac{\alpha \pi}{360} - \sin \dfrac{\alpha}{2} \cos \dfrac{\alpha}{2} \right) R^2}{\left(\dfrac{\alpha \pi}{360} - \sin \dfrac{\alpha}{2} \cos \dfrac{\alpha}{2} \right) R_1^2} = \frac{R^2}{R_1^2} = \frac{AB^2}{A_1 B_1^2}.$$

Applying this result to segments I and IV in Figure 4-8, we find:

$$\frac{\text{area IV}}{\text{area I}} = \frac{AB^2}{AD^2} = \frac{3AD^2}{AD^2} = 3.$$

$\therefore$ area IV = area I + area II + area III.

(b) Area trapezoid $ABCD$ = area segment on AB − (area I + area II + area III) = area segment on AB − area IV = area lune.

(c) See Exercise 2, part (f), in Exercises 4-2.

4   $m\angle CAB = \frac{1}{2}m$ arc $BC = \frac{1}{2}m$ arc AEB (since segment III $\sim$ segment IV).

EXERCISES 4-5

1   (a) $x = a\sqrt{2} \Rightarrow x^2 = 2a^2 ; \frac{a}{x} = \frac{x}{2a}$.   (b) $x = a\sqrt{2} \Rightarrow x^2 = a^2 + a^2$.

2   Eliminate y from $\frac{a}{x} = \frac{x}{y}$ and $\frac{a}{x} = \frac{y}{2a}$.

4   $(x^2 = ay$ and $y^2 = 2ax) \Rightarrow (x^2 y^2 = 2a^2 xy)$. $\therefore xy = 2a^2$.

5   $\frac{4}{x} = \frac{x}{y} = \frac{y}{8} \Rightarrow 4y = x^2$ and $8x = y^2$. Graph these equations using a unit of 1 inch. Apply the same method as in Figure 4-13.

EXERCISES 4-6

1   Let M be the midpoint of $\overline{PQ}$. $\overline{BM} \cong \overline{PM} \cong \overline{MQ} \cong \overline{OB}$ (M is the center of the circumcircle of the right triangle PBQ). Then use the facts that $\triangle BOM$ and $\triangle MBQ$ are isosceles and that the measure of an exterior angle of a triangle is equal to the sum of the measures of the opposite interior angles.

2   Construct an equilateral triangle to obtain an angle of 60°.

4   Since $GH < d$, the conchoid passes through G and has a loop at G.

6   (b) No.

9

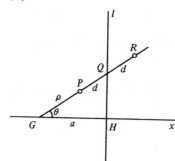

Let G be the origin, $\overrightarrow{GH}$ the positive x-axis, P and R points of the conchoid. Then $\rho_P = GP = GQ - PQ = GH \sec\theta - d = a \sec\theta - d$. $\rho_R = \rho_P + 2d = a \sec\theta + d$.

11   See the figure in the answer for Exercise 9. If x and y are the coordinates of P, then a and $\frac{ay}{x}$ are those of Q. The distance formula, applied to P and Q, gives $(x-a)^2 + (y - \frac{ay}{x})^2 = d^2$.

(b) $\rho = \sqrt{x^2 + y^2}$, $\sec\theta = \frac{\sqrt{x^2 + y^2}}{x}$. Substitution in

$\rho = a\sec\theta \pm d$ gives $\sqrt{x^2 + y^2} = \frac{a\sqrt{x^2 + y^2}}{x} \pm d$;

$(x - a)\sqrt{x^2 + y^2} = \pm\, dx$.

12   (a) Symmetry with respect to the line $\overleftrightarrow{GH}$ (the x-axis).
(b) If (θ_0, ρ_0) satisfies the equation, then also $(-\theta_0, \rho_0)$.
(c) If (x_0, y_0) satisfies the equation, then also $(x_0, -y_0)$.

EXERCISES 4-7

3   (a) See partial answer to Exercise 2 in Exercises 4-6.
(b) A given line segment $\overline{AB}$ can be trisected as follows: draw a ray $\overrightarrow{AC}$; lay off on $\overrightarrow{AC}$ congruent line segments $\overline{AD}$, $\overline{DE}$, and $\overline{EF}$; draw $\overline{FB}$ and lines through D and E parallel to $\overline{FB}$.

4    $75° = 60° + \frac{1}{2} \cdot 30°$.

6   See partial answer to part (b) of Exercise 3.

9   (a) $n \cdot \frac{1}{2} \cdot \frac{p}{n} \cdot r = \frac{1}{2}pr$.

10   The definition of the quadratrix gives: $\dfrac{\theta}{\frac{1}{2}\pi} = \dfrac{\rho\sin\theta}{r}$.

12   (b) $\lim\limits_{\theta\to 0}\rho = \lim\limits_{\theta\to 0}\dfrac{2r\theta}{\pi\sin\theta} = \lim\limits_{\theta\to 0}\dfrac{2r}{\pi}\cdot\lim\limits_{\theta\to 0}\dfrac{1}{\frac{\sin\theta}{\theta}} = \dfrac{2r}{\pi}\cdot 1$.

(c) $e = \lim\limits_{\theta\to 0}\rho$. Apply the result of part (b).

13   The assumption $\frac{q}{r} > \frac{r}{e}$ implies the existence of a number $a < e$ such that $\frac{q}{r} = \frac{r}{a}$. Since $a < e$ there is a segment $\overline{AP}$ of length a such that P is between A and E (see the figure).

Let $\overline{PR} \perp \overline{AB}$. We have again $\dfrac{q}{r} = \dfrac{l(\text{arc } PQ)}{a}$, and hence, $l(\text{arc } PQ) = r$.

$\dfrac{m\angle BAR}{m\angle BAD} = \dfrac{RP}{DA} \Rightarrow \dfrac{l(\text{arc } PF)}{l(\text{arc } PQ)} = \dfrac{RP}{DA}, \dfrac{l(\text{arc } PF)}{r} = \dfrac{RP}{r},$

l (arc PF) = RP. Hence, $\frac{1}{2}AP \cdot l$ (arc PF) = $\frac{1}{2}AP \cdot RP$, or area sector APF = area $\triangle APR$, which is not true.

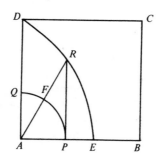

14 (a) Draw the quadratrix in a square of side 1. Using the same letters as in the text, we then have $\frac{q}{r} = \frac{r}{e}$. Hence, a line segment of length q can be constructed as the fourth proportional to e, r, and r. Since $4q = 2\pi r = 2\pi$ (since $r = 1$) we have $\pi = 2q$.

16 (a) Let E be the foot of the perpendicular from B to $\overline{FP}$ and R the foot of the perpendicular from P to $\overrightarrow{BD}$. Then the triangles BEF, BEP, and BRP are congruent.

EXERCISES 4-8

1 (a) $\frac{2}{5}$. (c) $-5, 5$. (e) $-2\frac{1}{2}, 2\frac{1}{2}$. (g) $-6, 1$. (i) $-3, -2$.
(k) $1, 2, 3$. (m) $-1, 1\frac{1}{2}, 2$.

2 (a) $\sqrt{5}$ is a root of the equation but not equal to $\frac{5}{1}$, $-\frac{5}{1}$, $\frac{1}{1}$, or $-\frac{1}{1}$.
(b) Use the equation $x^3 - 5 = 0$. (d) Use the equation $x^2 - p = 0$.

3 Since $\frac{1}{a} = \frac{b}{ab}$, a line segment of length ab is the fourth proportional to the segments of lengths 1, a, and b.

4 $\frac{a}{b}$ is the fourth proportional to b, 1, and a.

EXERCISES 5-2

1 Area square with side of length w is w^2. Area square with twice this area is $2w^2$. Length side of this square is $\sqrt{2w^2} = w\sqrt{2}$. Diagonal of square with side of length w has length $\sqrt{w^2 + w^2} = \sqrt{2w^2} = w\sqrt{2}$.

3

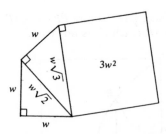

5   Construct an isosceles right triangle with legs of length r. Construct segments of lengths $r\sqrt{2}$ and $r\sqrt{3}$, as in Exercise 3. These segments and a segment of length $2r$ are the radii of the required circles.

7   If $\triangle ABC$ is isosceles.

9   Area square = 64. Area rectangle = 65. There is a "thin" parallelogram along the diagonal of the rectangle which is not covered. The angles at the lower left vertex are not complementary.

11   (a)   Vertex angles of regular hexagons measure 120°. Three hexagons with a common vertex fill the plane around that vertex.

      (b)   The vertex angle of a regular pentagon measures 108°. 360 is not a multiple of 108.

13   (a)   If the measures of nonoverlapping angles with a common vertex total 360°, the angles are coplanar.

      (b)   Three times the measures of the vertex angles of regular polygons are 180°, 270°, 324°, and equal to or greater than 360° for $n = 3, 4, 5$, and greater than 5.

EXERCISES 5-5

1

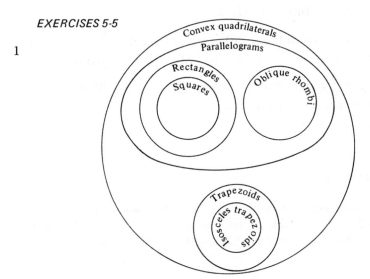

3

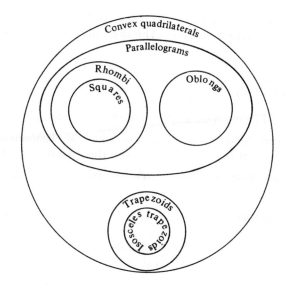

5

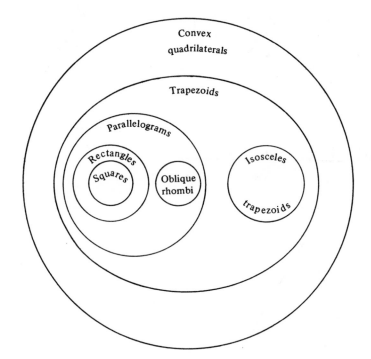

7   A square is a rectangular parallelogram whose sides are congruent.
9   An oblong is a rectangle whose adjacent sides are not congruent.

11   The motion of the edge of the hubcap is a combination of sliding,
     or translation, and rotation. The tire rolls on the road without
     slipping.

EXERCISES 6-3

1   Rectilineal angle, right angle, parallel straight lines.
3   If you have difficulty here, you can take consolation from the fact
    that great mathematicians have taken these as undefined terms.
5   Line, extremities of a line, extremities of a surface, plane angle.

EXERCISES 6-6

1   The compass is "set" anew for the second circle.

3   (a)   (b)

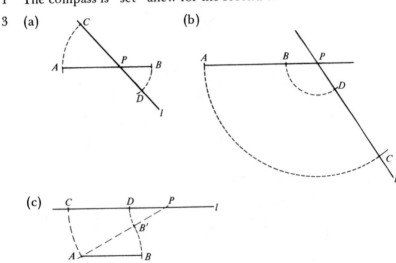

   (c)

   5

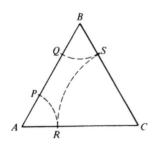

7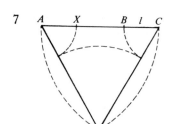

9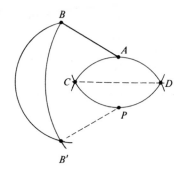

Draw C_1 $(P, \overline{PA})$, the circle with center at P and radius $\overline{PA}$. Draw C_2 $(A, \overline{AP})$. C_1 intersects C_2 at C and D. Draw C_3 $(D, \overline{DB})$ and C_4 $(C, \overline{CB})$ to intersect at B'. $\overline{PB'} \cong \overline{AB}$. The dotted lines are not needed or used.

EXERCISES 6-7

1 Consider the 1-1 correspondence between $\triangle ABC$ and $\triangle BAC$ given by $A \leftrightarrow B$, $B \leftrightarrow A$, and $C \leftrightarrow C$. Then $\overline{AB} \cong \overline{BA}$, $\overline{BC} \cong \overline{AC}$, $\overline{AC} \cong \overline{BC}$. $\therefore \triangle ABC \cong \triangle BAC$. $\therefore \angle A \cong \angle B$.

3 (a)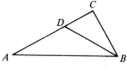

Assume D lies between A and C. Then $\overline{AD} < \overline{AC}$. This contradicts the given condition $\overline{AC} \cong \overline{AD}$. $\therefore D$ does not lie between A and C.

5 Determine B and C as before. Determine D as the intersection of circles C_1 (B, r) and C_2 (C, r), where $r > \frac{1}{2} BC$. The proof is unchanged.

7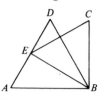

Construct equilateral $\triangle ABD$. Construct the midpoint E of AD. Construct equilateral $\triangle EBC$.

9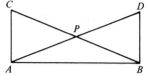

Since all right angles are congruent, $\triangle ABD \cong \triangle BAC$ by SAS. $\therefore \overline{AD} \cong \overline{BC}$.

11 (a) $\angle CAB \cong \angle DBA$; $\angle DAB \cong \angle CBA$. $\therefore \angle CAD \cong \angle DBC$ (Axiom III).

(b) $\overline{AP} \cong \overline{BP}$ by Proposition 6 applied to $\triangle ABP$.
$\overline{CP} \cong \overline{DP}$ by Axiom III.

EXERCISES 6-8

1 Prove that the vertical angle of $\angle ACD$ is greater than $\angle ABC$.

3 An indirect proof can be given by applying Propositions 6 and 18.

5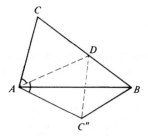

(1) $\overline{AP} + \overline{PB} < \overline{AP} + (\overline{PD} + \overline{DB})$
(Proposition 20)
$= \overline{AD} + \overline{DB} < (\overline{AC} + \overline{CD}) + \overline{DB}$
$= \overline{AC} + \overline{CB}$.
(2) $\angle APB > \angle ADB$ (Proposition 16)
$> \angle C$.

7 No.

9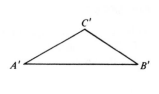

Let $\triangle ABC'' \cong \triangle A'B'C'$, and let $\overline{AD}$ bisect $\angle CAC''$. Then
$\triangle ADC \cong \triangle ADC''$ (Proposition 4). $\therefore \overline{CD} \cong \overline{C''D}$; $\overline{CB} = \overline{CD} + \overline{DB}$
$= \overline{C''D} + \overline{DB} > \overline{C''B} \cong \overline{C'B'}$.

11 $\angle ADC > \angle B$ (Proposition 16) $\cong \angle C$ (Proposition 6).
$\therefore \overline{AC} > \overline{AD}$ (Proposition 19).

12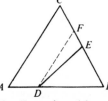

Suppose $\overline{BE} < \overline{BD}$. Draw $\overline{DF} \parallel \overline{AC}$, and
apply Exercise 11 to $\overline{DE}$ in the equilateral
$\triangle DBF$.

13 Use Exercises 11 and 12 to prove that each of $\overline{AD}$ and $\overline{DE}$ is less
than a side.

15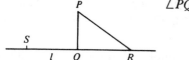

$\angle PQR \cong \angle PQS > \angle PRQ. \quad \therefore \overline{PR} > \overline{PQ}.$

EXERCISES 6-9

1

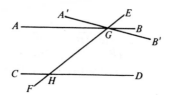

(a) If $\angle EGB \cong \angle FHC$, then $\angle EGB \cong \angle GHD$ since $\angle FHC \cong \angle GHD$ by Proposition 15. $\therefore \overleftrightarrow{AB} \parallel \overleftrightarrow{CD}$ by Proposition 28a.

(b) $\angle EGB + \angle BGH \cong 2$ right angles. By substitution,
   $\angle GHD + \angle BGH \cong 2$ right angles. $\therefore \overleftrightarrow{AB} \parallel \overleftrightarrow{CD}$.

3 See the figure in the answer to Exercise 1. $\angle AGF \cong \angle DHE$ by Proposition 29a. $\angle EGB \cong \angle AGF$ (Proposition 15).
   $\therefore \angle EGB \cong \angle DHE$.

5 See the figure in the answer to Exercise 1.
   (a) Given: $\overleftrightarrow{HD} \parallel \overleftrightarrow{GA}$. Construct $\angle HGA' \cong \angle GHD$. Then $\overleftrightarrow{GA'} \parallel \overleftrightarrow{HD}$ (Proposition 27) and hence $\overleftrightarrow{GA'} = \overleftrightarrow{GA}$ (otherwise Proposition B is contradicted). $\therefore \angle GHD \cong \angle HGA$.
   (b) $\angle AGH + \angle BGH = 2$ right angles. $\angle AGH \cong \angle GHD$.
   $\therefore \angle GHD + \angle BGH = 2$ right angles.
   (c) $\angle BGH + \angle GHD = 2$ right angles, because $\overleftrightarrow{AB} \parallel \overleftrightarrow{CD}$.
   Let $\overleftrightarrow{A'GB'}$ be a line such that $\angle B'GH + \angle GHD < 2$ right angles.
   Then $\angle B'GH + \angle GHD < \angle BGH + \angle GHD$, and, by subtraction,
   $\angle B'GH < \angle BGH$. Further, $\overleftrightarrow{A'GB'} \nparallel \overleftrightarrow{CD}$ by Proposition B.
   $\therefore \overleftrightarrow{A'GB'}$ meets $\overleftrightarrow{CD}$ on the B' side of GH. (A complete modern proof of this would require definitions and postulates on "betweenness" not recognized by Euclid.)

7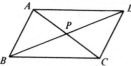

   $\triangle ABD \cong \triangle CDB$ by Proposition 26
   $\therefore \overline{AB} \cong \overline{CD}, \overline{AD} \cong \overline{CB}, \angle A \cong \angle C$,
   and $\angle ABC \cong \angle CDA$ (Axiom II).

9 Using the figure for Exercise 7, $\triangle APB \cong \triangle CPD$ (ASA).
   $\therefore \overline{PA} \cong \overline{PC}$.

11   In the figure for Exercise 7, if $\overline{BP} \cong \overline{PD}$ and $\overline{AP} \cong \overline{PC}$, then
      $\triangle APB \cong \triangle CPD$, $\therefore \angle ABP \cong \angle CDP$ and $\overline{AB} \cong \overline{CD}$. $\therefore$ $ABCD$ is a
      parallelogram by Proposition 33.

13   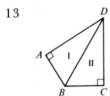 Any counter example will do; e.g., let I and II be
      two non-isosceles congruent right triangles fitted
      together to form a quadrilateral with $\angle A \cong \angle C$.
      Since $\angle ABD \not\cong \angle CDB$, $\overline{AB} \nparallel \overline{DC}$.

15   Extend $\overline{AB}$ and $\overline{DC}$ to meet at E. $\angle A \cong \angle D$,
      $\therefore$ $\overline{AE} \cong \overline{DE}$. $\angle EBC \cong \angle ECB$, $\overline{BE} \cong \overline{CE}$.
      $\therefore$ by subtraction, $\overline{AB} \cong \overline{DC}$.

17   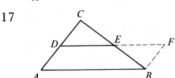 $\triangle DCE \cong \triangle FBE$ by SAS.
      $\overline{AD} \cong \overline{DC} \cong \overline{BF}$. $\angle DCE \cong \angle FBE$.
      $\therefore$ $ADFB$ is a parallelogram and
      $\therefore$ $\overline{DE} \parallel \overline{AB}$.

EXERCISES 6-10

1   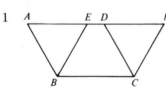 $\triangle ABE \cong \triangle DCF$
      $EBCD \cong EBCD$
      $\therefore$ $Area \square ABCD = area \square EBCF$.

3   Use Proposition 37 and indirect reasoning.

5.   Apply Proposition 38.

7.   Construct a triangle with same height and triple the base (or same
      base and triple height).

9.   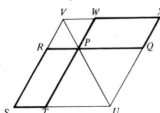 On $\overline{PQ}$ extended construct
      $\square PRST \cong \square ABCD$. Extend $\overline{ST}$,
      $\overline{SR}$, and construct $\overline{QU} \parallel \overline{PT}$ to meet
      $\overleftrightarrow{ST}$ at U. $\overrightarrow{UP}$ intersects $\overleftrightarrow{SR}$ at V.
      $\square PQXW = \square PRST$ in area.

EXERCISES 6-11

1 Draw $\overline{BK}$ and $\overline{AE}$. $ECML = 2\,(\triangle ECA)$ in area. $ACKH = 2\,(\triangle BCK)$ in area. $\triangle ECA \cong \triangle BCK$. $\therefore ECML = ACKH$ in area.

3 Let P and Q be the points of intersection of $\overline{AB}$ and $\overline{FC}$ and of $\overline{AD}$ and $\overline{FC}$. $\angle FPB \cong \angle APQ$, $\angle PFB \cong \angle PAQ$, $\therefore \angle AQP \cong \angle FBP \cong$ right angle.

5 Let I and J be the points of intersection of $\overleftrightarrow{HC}$ with $\overline{AB}$ and $\overline{MN}$. Let K and L be the points of intersection of $\overrightarrow{MA}$ with $\overline{DE}$ and $\overrightarrow{NB}$ with $\overline{FG}$. $\overline{AK} \cong \overline{HC} \cong \overline{BL} \cong \overline{AM} \cong \overline{BN}$. In area, $\square ADEC = \square AKHC = \square MAIJ$ and $\square BFGC = \square BLHC = \square NBIJ$. The theorem follows by addition of equal areas.

7 Take $m\angle C = 90°$ and parallelograms $ACED$ and $CBFG$ to be squares on $\overline{AC}$ and $\overline{CB}$.

EXERCISES 6-13

1 (a) Let $AB = a + b$. Construct the square $ABCD$ on $\overline{AB}$ and draw its diagonal $\overline{AC}$. Take $\overline{EF} \parallel \overline{AD}$, $\overline{GH} \parallel \overline{AB}$ at distance a from $\overline{AD}$ and $\overline{AB}$. $\overline{EF}$ and $\overline{GH}$ intersect on $\overline{AC}$ at P, forming two squares and two congruent rectangles.

(b) If a line is divided into two parts, the square on the line equals the sum of the squares on the parts plus twice the rectangle on the parts.

3

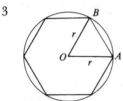

Since $m\angle AOB = 60°$ and $\triangle AOB$ is isosceles, the triangle is equilateral and $AB = r$. The opposite sides of squares and hexagons are parallel. $\therefore$ Wrenches with parallel jaws fit.

5 If a 9-gon could be constructed, then angles of $40°$ and $20°$ $(= \frac{1}{2} \cdot 40°)$ could be constructed. But it is impossible to construct $\frac{1}{3} \cdot 60° = 20°$.

EXERCISES 6-14

1 $496 = 1 + 2 + 4 + 8 + 16 + 31 + 62 + 124 + 248$.

3 (a) 15. (b) 26. (c) 12. (d) 12.

EXERCISES 7-2

1   (a) 2, 3, 5, 7, 11, 13, 17, 19, 23, 29, 31, 37, 41, 43, 47, 53, 59, 61, 67, 71, 73, 79, 83, 89, 97.   (c) The largest prime less than $\sqrt{n}$.

2

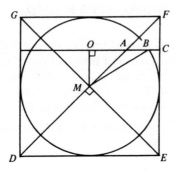

(a) The figure shows a cross section of cone, sphere, and cylinder, by a plane containing the axis, line $\overleftrightarrow{MO}$, of the cylinder. $\overline{OA}$, $\overline{OB}$, and $\overline{OC}$ are the radii of the circular regions which are cut off from plane P by cone, sphere, and cylinder. Let these radii be $r_1, r_2,$ and r_3. $m\angle OMA = 45° \Rightarrow OM = OA = r_1.$ $MB = \frac{1}{2}GF = OC = r_3.$ $OB^2 = MB^2 - OM^2$, or $r_2^2 = r_3^2 - r_1^2.$ $\therefore \pi r_2^2 = \pi r_3^2 - \pi r_1^2.$

(b) Let P' be a plane parallel to P at a distance $\frac{2r_3}{n}$ (n a large integer). Then the parts of cone, sphere, and cylinder (including their interiors) between P and P' can be considered to be thin disks with volumes $\pi r_1^2 \cdot \frac{2r_3}{n}$, $\pi r_2^2 \cdot \frac{2r_3}{n}$, and $\pi r_3^2 \cdot \frac{2r_3}{n}$. We then have $\pi r_2^2 \cdot \frac{2r_3}{n} = \pi r_3^2 \cdot \frac{2r_3}{n} - \pi r_1^2 \cdot \frac{2r_3}{n}$. Consider cone, sphere, and cylinder each as a stack of such thin disks. Then the volume of the stack of disks in the sphere is equal to the volume of the stack of disks in the cylinder minus the volume of the stack of disks in the cone.

(c) Volume sphere $= \frac{4}{3}\pi r_3^3$, volume cylinder $= \pi r_3^2 \cdot 2r_3$, volume cone $= 2 \cdot \frac{1}{3} \cdot \pi r_3^2 \cdot r_3;$ $\frac{4}{3}\pi r_3^3 = 2\pi r_3^3 - \frac{2}{3}\pi r_3^3.$

3   (a) Equation: $\rho = \theta$.

(c) Place the given angle with its vertex in O, one side along the positive x-axis, and the other side in the first or second quadrant. Let P be the point of intersection of the other side and the spiral and P' the point of the spiral such that $OP' = \frac{1}{3}OP$. Then the angle formed by $\overrightarrow{OP'}$ and the positive x-axis is $\frac{1}{3}$ of the given angle.

EXERCISES 7-3

2 Draw the perpendicular bisectors of two sides of $\triangle ABC$.

3 Bisect two angles or one angle and an exterior angle of the triangle formed by the three lines.

5 If the lines are concurrent or all three parallel. No.

7 (a) Let $\overline{CD}$ be an altitude of $\triangle ABC$ (assume D between M and B). Then we have

$$CD^2 = m^2 - MD^2 \qquad (1)$$
$$CD^2 = b^2 - (MD + \tfrac{c}{2})^2 \qquad (2)$$
$$CD^2 = a^2 - (\tfrac{c}{2} - MD)^2 \qquad (3)$$

Eliminate CD and MD, for instance by subtracting 2 times equation (1) from the sum of the equations (2) and (3).

(b) Use the fact that $m = \dfrac{c}{2}$.

EXERCISES 7-4

2 Let $\overline{CD}$ be an altitude of $\triangle ABC$ (assume D between A and B); let $CD = h$ and $AD = x$. Then $h^2 = b^2 - x^2$ and $h^2 = a^2 - (c-x)^2$, and hence $b^2 - x^2 = a^2 - (c-x)^2$. Solving for x, we get $x = \dfrac{b^2 + c^2 - a^2}{2c}$. Hence, $h^2 (= b^2 - x^2) = b^2 - \left(\dfrac{b^2 + c^2 - a^2}{2c}\right)^2$

$$= \left(b + \frac{b^2 + c^2 - a^2}{2c}\right)\left(b - \frac{b^2 + c^2 - a^2}{2c}\right)$$

$$= \frac{(b+c)^2 - a^2}{2c} \cdot \frac{a^2 - (b-c)^2}{2c} = \frac{2s \cdot (2s-2a)(2s-2c)(2s-2b)}{4c^2}$$

$$= \frac{4}{c^2} \cdot s(s-a)(s-b)(s-c). \quad \therefore \text{ area } \triangle ABC = \tfrac{1}{2}c \cdot h$$

$$= \tfrac{1}{2}c \sqrt{\frac{4}{c^2} \cdot s(s-a)(s-b)(s-c)} = \sqrt{s(s-a)(s-b)(s-c)}.$$

3 8, 12.

4 3, 7.

6 (a) $60x^2 + 2520$. (b) $x^4 - 12x^2 + 36$.
   (c) $3x^5 - 4x^4 + 13x^2 + 45x - 228$.

7 (a) $\varsigma \overline{\lambda\beta} \stackrel{\circ}{\mathrm{M}} \overline{\theta}$. (b) $\Delta^{\mathrm{Y}} \overline{\mu\alpha} \wedge \stackrel{\circ}{\mathrm{M}} \overline{\gamma}$. (c) $\mathrm{K}^{\mathrm{Y}} \overline{\beta} \stackrel{\circ}{\mathrm{M}} \overline{\mu\eta} \wedge \Delta^{\mathrm{Y}} \overline{\gamma}$.

EXERCISES 7-5

2 (a) $2\pi^2 r^2 R$. (b) $\dfrac{3\pi^2}{8}$. (c) $4\pi^2 r R$. (d) $\dfrac{3\pi^2}{2}$.

5 (a) $DE \cdot DO = DC^2$, $DE \cdot \dfrac{AC+BC}{2} = AC \cdot BC$, $DE = \dfrac{2AC \cdot BC}{AC + BC}$.

6 Prove that $(AE + EC) \cdot BD = AB \cdot CD + AD \cdot BC$.

7 The central angle subtended by the chord of length k is of measure 2α. Bisect this angle.

EXERCISES 8-1

1 (a) MMDCCLXVII   (c) MCDXXIII

2 (a) MCXI

EXERCISES 8-4

1 (a)   (c)

2 (a) 36146   (c) 8760

3 (a) 7239   (c) No.

4 (c) Yes.

EXERCISES 8-6

1 (a) 5. (c) 21. (e) 380.

2 (a) 24_5. (c) 31_5. (e) 100_5.

3 (a) 34003_5. (c) 222333_5.

4 (a) 5_6. (c) 40_6. (e) 104_6. (g) 1304_6.

5 (a) 6611. (b) 37.

6 $1 \times 52_{12}$.

7 (a) 101_2. (c) 10110_2. (e) 1010001_2.

8 105_8.

9 (a) 9.

EXERCISES 8-7

1 (a) 1055_6. (c) 5120_6.

2 (a) 1405_8. (c) 6240_8.

3 (a) 202_5. (c) 401_5.

4 (a) 106_7. (c) 41643_7.

EXERCISES 8-8

3 (a) 2132221_4. (c) 20100330_4.

4 (a) 14616_7. (c) 14044656_7.

5 (a) 11000110_2. (c) 100111101010_2.

6 (a) 123_4. (c) 123_4.

7 (a) 542_7. (c) 345_7.

EXERCISES 8-9

1 (a) 261.25. (c) $261.008\overline{3}$. (e) 0.1. (g) 261.25°.

2 (a) 8;45. (c) 4,1;10.

3 (a) 0;7,30. (b) 133° 53′ 55″. (c) 25;19,34,52,30.
   (d) 20° 10′ 31″ 8‴ 12⁗.

4 (a) 0;30. (c) 0;12. (e) 0;6. (g) 0;32. (i) 0;26.

5 (a) 0;7,30. (c) 0;3,20. (e) 0;2,24. (g) 0;1,52,30.

6 (a) $0;\overline{5,27,16,21,49}$ (c) $0;4,\overline{17,8,34}$.

8 (a) $\frac{1}{2}, \frac{1}{4}, \frac{1}{5}, \frac{1}{8}$. (c) $\frac{1}{2}, \frac{1}{3}, \frac{1}{4}, \frac{1}{6}, \frac{1}{8}, \frac{1}{9}$.

11 (c) F. I.
    91 10 1

13 6° 16′ 54″ 11‴ 15⁗.

EXERCISES 8-10

2 (a) $0.1\overline{6}$. (c) $0.\overline{285714}$. (e) $0.5\overline{4}$. (g) $0.\overline{076923}$. (i) 1.4.

4 (a) $\frac{14}{99}$. (c) $\frac{1}{2}$.

EXERCISES 8-11

2 (a) 2. (c) 4. (e) 0.

3   (a) 1.   (c) 0.   (e) 2.
4   (a) 3.   (c) 0.   (e) 3.
5   The number 2 has no multiplicative inverse.

EXERCISES 8-12

1   (c) 4×10^{12}. (d) 4×10^{18}.
2   (b) 8.225×10^{18}.
3   (a) 0.000,000,018. Eighteen billionths.

INDEX

A CATALOG OF SELECTED
DOVER BOOKS
IN ALL FIELDS OF INTEREST

A CATALOG OF SELECTED DOVER
BOOKS IN ALL FIELDS OF INTEREST

CONCERNING THE SPIRITUAL IN ART, Wassily Kandinsky. Pioneering work by father of abstract art. Thoughts on color theory, nature of art. Analysis of earlier masters. 12 illustrations. 80pp. of text. 5⅜ x 8½.   23411-8 Pa. $4.95

ANIMALS: 1,419 Copyright-Free Illustrations of Mammals, Birds, Fish, Insects, etc., Jim Harter (ed.). Clear wood engravings present, in extremely lifelike poses, over 1,000 species of animals. One of the most extensive pictorial sourcebooks of its kind. Captions. Index. 284pp. 9 x 12.   23766-4 Pa. $14.95

CELTIC ART: The Methods of Construction, George Bain. Simple geometric techniques for making Celtic interlacements, spirals, Kells-type initials, animals, humans, etc. Over 500 illustrations. 160pp. 9 x 12. (USO)   22923-8 Pa. $9.95

AN ATLAS OF ANATOMY FOR ARTISTS, Fritz Schider. Most thorough reference work on art anatomy in the world. Hundreds of illustrations, including selections from works by Vesalius, Leonardo, Goya, Ingres, Michelangelo, others. 593 illustrations. 192pp. 7⅛ x 10¼.   20241-0 Pa. $9.95

CELTIC HAND STROKE-BY-STROKE (Irish Half-Uncial from "The Book of Kells"): An Arthur Baker Calligraphy Manual, Arthur Baker. Complete guide to creating each letter of the alphabet in distinctive Celtic manner. Covers hand position, strokes, pens, inks, paper, more. Illustrated. 48pp. 8¼ x 11.   24336-2 Pa. $3.95

EASY ORIGAMI, John Montroll. Charming collection of 32 projects (hat, cup, pelican, piano, swan, many more) specially designed for the novice origami hobbyist. Clearly illustrated easy-to-follow instructions insure that even beginning papercrafters will achieve successful results. 48pp. 8¼ x 11.   27298-2 Pa. $3.50

THE COMPLETE BOOK OF BIRDHOUSE CONSTRUCTION FOR WOODWORKERS, Scott D. Campbell. Detailed instructions, illustrations, tables. Also data on bird habitat and instinct patterns. Bibliography. 3 tables. 63 illustrations in 15 figures. 48pp. 5¼ x 8½.   24407-5 Pa. $2.50

BLOOMINGDALE'S ILLUSTRATED 1886 CATALOG: Fashions, Dry Goods and Housewares, Bloomingdale Brothers. Famed merchants' extremely rare catalog depicting about 1,700 products: clothing, housewares, firearms, dry goods, jewelry, more. Invaluable for dating, identifying vintage items. Also, copyright-free graphics for artists, designers. Co-published with Henry Ford Museum & Greenfield Village. 160pp. 8¼ x 11.   25780-0 Pa. $10.95

HISTORIC COSTUME IN PICTURES, Braun & Schneider. Over 1,450 costumed figures in clearly detailed engravings–from dawn of civilization to end of 19th century. Captions. Many folk costumes. 256pp. 8⅜ x 11¾.   23150-X Pa. $12.95

STICKLEY CRAFTSMAN FURNITURE CATALOGS, Gustav Stickley and L. & J. G. Stickley. Beautiful, functional furniture in two authentic catalogs from 1910. 594 illustrations, including 277 photos, show settles, rockers, armchairs, reclining chairs, bookcases, desks, tables. 183pp. 6½ x 9¼. 23838-5 Pa. $11.95

AMERICAN LOCOMOTIVES IN HISTORIC PHOTOGRAPHS: 1858 to 1949, Ron Ziel (ed.). A rare collection of 126 meticulously detailed official photographs, called "builder portraits," of American locomotives that majestically chronicle the rise of steam locomotive power in America. Introduction. Detailed captions. xi + 129pp. 9 x 12. 27393-8 Pa. $13.95

AMERICA'S LIGHTHOUSES: An Illustrated History, Francis Ross Holland, Jr. Delightfully written, profusely illustrated fact-filled survey of over 200 American lighthouses since 1716. History, anecdotes, technological advances, more. 240pp. 8 x 10¾. 25576-X Pa. $12.95

TOWARDS A NEW ARCHITECTURE, Le Corbusier. Pioneering manifesto by founder of "International School." Technical and aesthetic theories, views of industry, economics, relation of form to function, "mass-production split" and much more. Profusely illustrated. 320pp. 6⅛ x 9¼. (USO) 25023-7 Pa. $9.95

HOW THE OTHER HALF LIVES, Jacob Riis. Famous journalistic record, exposing poverty and degradation of New York slums around 1900, by major social reformer. 100 striking and influential photographs. 233pp. 10 x 7⅞. 22012-5 Pa. $11.95

FRUIT KEY AND TWIG KEY TO TREES AND SHRUBS, William M. Harlow. One of the handiest and most widely used identification aids. Fruit key covers 120 deciduous and evergreen species; twig key 160 deciduous species. Easily used. Over 300 photographs. 126pp. 5⅜ x 8½. 20511-8 Pa. $3.95

COMMON BIRD SONGS, Dr. Donald J. Borror. Songs of 60 most common U.S. birds: robins, sparrows, cardinals, bluejays, finches, more—arranged in order of increasing complexity. Up to 9 variations of songs of each species. Cassette and manual 99911-4 $8.95

ORCHIDS AS HOUSE PLANTS, Rebecca Tyson Northen. Grow cattleyas and many other kinds of orchids—in a window, in a case, or under artificial light. 63 illustrations. 148pp. 5⅜ x 8½. 23261-1 Pa. $5.95

MONSTER MAZES, Dave Phillips. Masterful mazes at four levels of difficulty. Avoid deadly perils and evil creatures to find magical treasures. Solutions for all 32 exciting illustrated puzzles. 48pp. 8¼ x 11. 26005-4 Pa. $2.95

MOZART'S DON GIOVANNI (DOVER OPERA LIBRETTO SERIES), Wolfgang Amadeus Mozart. Introduced and translated by Ellen H. Bleiler. Standard Italian libretto, with complete English translation. Convenient and thoroughly portable—an ideal companion for reading along with a recording or the performance itself. Introduction. List of characters. Plot summary. 121pp. 5¼ x 8½. 24944-1 Pa. $3.95

TECHNICAL MANUAL AND DICTIONARY OF CLASSICAL BALLET, Gail Grant. Defines, explains, comments on steps, movements, poses and concepts. 15-page pictorial section. Basic book for student, viewer. 127pp. 5⅜ x 8½. 21843-0 Pa. $4.95

THE CLARINET AND CLARINET PLAYING, David Pino. Lively, comprehensive work features suggestions about technique, musicianship, and musical interpretation, as well as guidelines for teaching, making your own reeds, and preparing for public performance. Includes an intriguing look at clarinet history. "A godsend," The Clarinet, Journal of the International Clarinet Society. Appendixes. 7 illus. 320pp. 5⅜ x 8½. 40270-3 Pa. $9.95

HOLLYWOOD GLAMOR PORTRAITS, John Kobal (ed.). 145 photos from 1926-49. Harlow, Gable, Bogart, Bacall; 94 stars in all. Full background on photographers, technical aspects. 160pp. 8⅜ x 11¼. 23352-9 Pa. $12.95

THE ANNOTATED CASEY AT THE BAT: A Collection of Ballads about the Mighty Casey/Third, Revised Edition, Martin Gardner (ed.). Amusing sequels and parodies of one of America's best-loved poems: Casey's Revenge, Why Casey Whiffed, Casey's Sister at the Bat, others. 256pp. 5⅜ x 8½. 28598-7 Pa. $8.95

THE RAVEN AND OTHER FAVORITE POEMS, Edgar Allan Poe. Over 40 of the author's most memorable poems: "The Bells," "Ulalume," "Israfel," "To Helen," "The Conqueror Worm," "Eldorado," "Annabel Lee," many more. Alphabetic lists of titles and first lines. 64pp. 5$\frac{3}{16}$ x 8¼. 26685-0 Pa. $1.00

PERSONAL MEMOIRS OF U. S. GRANT, Ulysses Simpson Grant. Intelligent, deeply moving firsthand account of Civil War campaigns, considered by many the finest military memoirs ever written. Includes letters, historic photographs, maps and more. 528pp. 6⅛ x 9¼. 28587-1 Pa. $12.95

ANCIENT EGYPTIAN MATERIALS AND INDUSTRIES, A. Lucas and J. Harris. Fascinating, comprehensive, thoroughly documented text describes this ancient civilization's vast resources and the processes that incorporated them in daily life, including the use of animal products, building materials, cosmetics, perfumes and incense, fibers, glazed ware, glass and its manufacture, materials used in the mummification process, and much more. 544pp. 6⅛ x 9¼. (USO) 40446-3 Pa. $16.95

RUSSIAN STORIES/PYCCKNE PACCKA3bl: A Dual-Language Book, edited by Gleb Struve. Twelve tales by such masters as Chekhov, Tolstoy, Dostoevsky, Pushkin, others. Excellent word-for-word English translations on facing pages, plus teaching and study aids, Russian/English vocabulary, biographical/critical introductions, more. 416pp. 5⅜ x 8½. 26244-8 Pa. $9.95

PHILADELPHIA THEN AND NOW: 60 Sites Photographed in the Past and Present, Kenneth Finkel and Susan Oyama. Rare photographs of City Hall, Logan Square, Independence Hall, Betsy Ross House, other landmarks juxtaposed with contemporary views. Captures changing face of historic city. Introduction. Captions. 128pp. 8¼ x 11. 25790-8 Pa. $9.95

AIA ARCHITECTURAL GUIDE TO NASSAU AND SUFFOLK COUNTIES, LONG ISLAND, The American Institute of Architects, Long Island Chapter, and the Society for the Preservation of Long Island Antiquities. Comprehensive, well-researched and generously illustrated volume brings to life over three centuries of Long Island's great architectural heritage. More than 240 photographs with authoritative, extensively detailed captions. 176pp. 8¼ x 11. 26946-9 Pa. $14.95

NORTH AMERICAN INDIAN LIFE: Customs and Traditions of 23 Tribes, Elsie Clews Parsons (ed.). 27 fictionalized essays by noted anthropologists examine religion, customs, government, additional facets of life among the Winnebago, Crow, Zuni, Eskimo, other tribes. 480pp. 6⅛ x 9¼. 27377-6 Pa. $10.95

FRANK LLOYD WRIGHT'S DANA HOUSE, Donald Hoffmann. Pictorial essay of residential masterpiece with over 160 interior and exterior photos, plans, elevations, sketches and studies. 128pp. 9¼ x 10¾. 29120-0 Pa. $12.95

THE MALE AND FEMALE FIGURE IN MOTION: 60 Classic Photographic Sequences, Eadweard Muybridge. 60 true-action photographs of men and women walking, running, climbing, bending, turning, etc., reproduced from rare 19th-century masterpiece. vi + 121pp. 9 x 12. 24745-7 Pa. $10.95

1001 QUESTIONS ANSWERED ABOUT THE SEASHORE, N. J. Berrill and Jacquelyn Berrill. Queries answered about dolphins, sea snails, sponges, starfish, fishes, shore birds, many others. Covers appearance, breeding, growth, feeding, much more. 305pp. 5¼ x 8¼. 23366-9 Pa. $9.95

ATTRACTING BIRDS TO YOUR YARD, William J. Weber. Easy-to-follow guide offers advice on how to attract the greatest diversity of birds: birdhouses, feeders, water and waterers, much more. 96pp. 5³⁄₁₆ x 8¼. 28927-3 Pa. $2.50

MEDICINAL AND OTHER USES OF NORTH AMERICAN PLANTS: A Historical Survey with Special Reference to the Eastern Indian Tribes, Charlotte Erichsen-Brown. Chronological historical citations document 500 years of usage of plants, trees, shrubs native to eastern Canada, northeastern U.S. Also complete identifying information. 343 illustrations. 544pp. 6½ x 9¼. 25951-X Pa. $12.95

STORYBOOK MAZES, Dave Phillips. 23 stories and mazes on two-page spreads: Wizard of Oz, Treasure Island, Robin Hood, etc. Solutions. 64pp. 8¼ x 11. 23628-5 Pa. $2.95

AMERICAN NEGRO SONGS: 230 Folk Songs and Spirituals, Religious and Secular, John W. Work. This authoritative study traces the African influences of songs sung and played by black Americans at work, in church, and as entertainment. The author discusses the lyric significance of such songs as "Swing Low, Sweet Chariot," "John Henry," and others and offers the words and music for 230 songs. Bibliography. Index of Song Titles. 272pp. 6½ x 9¼. 40271-1 Pa. $9.95

MOVIE-STAR PORTRAITS OF THE FORTIES, John Kobal (ed.). 163 glamor, studio photos of 106 stars of the 1940s: Rita Hayworth, Ava Gardner, Marlon Brando, Clark Gable, many more. 176pp. 8⅜ x 11¼. 23546-7 Pa. $14.95

BENCHLEY LOST AND FOUND, Robert Benchley. Finest humor from early 30s, about pet peeves, child psychologists, post office and others. Mostly unavailable elsewhere. 73 illustrations by Peter Arno and others. 183pp. 5⅜ x 8½. 22410-4 Pa. $6.95

YEKL and THE IMPORTED BRIDEGROOM AND OTHER STORIES OF YIDDISH NEW YORK, Abraham Cahan. Film Hester Street based on Yekl (1896). Novel, other stories among first about Jewish immigrants on N.Y.'s East Side. 240pp. 5⅜ x 8½. 22427-9 Pa. $6.95

SELECTED POEMS, Walt Whitman. Generous sampling from *Leaves of Grass*. Twenty-four poems include "I Hear America Singing," "Song of the Open Road," "I Sing the Body Electric," "When Lilacs Last in the Dooryard Bloom'd," "O Captain! My Captain!"—all reprinted from an authoritative edition. Lists of titles and first lines. 128pp. 5³⁄₁₆ x 8¼. 26878-0 Pa. $1.00

CATALOG OF DOVER BOOKS

THE BEST TALES OF HOFFMANN, E. T. A. Hoffmann. 10 of Hoffmann's most important stories: "Nutcracker and the King of Mice," "The Golden Flowerpot," etc. 458pp. 5⅜ x 8½. 21793-0 Pa. $9.95

FROM FETISH TO GOD IN ANCIENT EGYPT, E. A. Wallis Budge. Rich detailed survey of Egyptian conception of "God" and gods, magic, cult of animals, Osiris, more. Also, superb English translations of hymns and legends. 240 illustrations. 545pp. 5⅜ x 8½. 25803-3 Pa. $13.95

FRENCH STORIES/CONTES FRANÇAIS: A Dual-Language Book, Wallace Fowlie. Ten stories by French masters, Voltaire to Camus: "Micromegas" by Voltaire; "The Atheist's Mass" by Balzac; "Minuet" by de Maupassant; "The Guest" by Camus, six more. Excellent English translations on facing pages. Also French-English vocabulary list, exercises, more. 352pp. 5⅜ x 8½. 26443-2 Pa. $9.95

CHICAGO AT THE TURN OF THE CENTURY IN PHOTOGRAPHS: 122 Historic Views from the Collections of the Chicago Historical Society, Larry A. Viskochil. Rare large-format prints offer detailed views of City Hall, State Street, the Loop, Hull House, Union Station, many other landmarks, circa 1904-1913. Introduction. Captions. Maps. 144pp. 9⅜ x 12¼. 24656-6 Pa. $12.95

OLD BROOKLYN IN EARLY PHOTOGRAPHS, 1865-1929, William Lee Younger. Luna Park, Gravesend race track, construction of Grand Army Plaza, moving of Hotel Brighton, etc. 157 previously unpublished photographs. 165pp. 8⅞ x 11¾. 23587-4 Pa. $13.95

THE MYTHS OF THE NORTH AMERICAN INDIANS, Lewis Spence. Rich anthology of the myths and legends of the Algonquins, Iroquois, Pawnees and Sioux, prefaced by an extensive historical and ethnological commentary. 36 illustrations. 480pp. 5⅜ x 8½. 25967-6 Pa. $10.95

AN ENCYCLOPEDIA OF BATTLES: Accounts of Over 1,560 Battles from 1479 B.C. to the Present, David Eggenberger. Essential details of every major battle in recorded history from the first battle of Megiddo in 1479 B.C. to Grenada in 1984. List of Battle Maps. New Appendix covering the years 1967-1984. Index. 99 illustrations. 544pp. 6½ x 9¼. 24913-1 Pa. $16.95

SAILING ALONE AROUND THE WORLD, Captain Joshua Slocum. First man to sail around the world, alone, in small boat. One of great feats of seamanship told in delightful manner. 67 illustrations. 294pp. 5⅜ x 8½. 20326-3 Pa. $6.95

ANARCHISM AND OTHER ESSAYS, Emma Goldman. Powerful, penetrating, prophetic essays on direct action, role of minorities, prison reform, puritan hypocrisy, violence, etc. 271pp. 5⅜ x 8½. 22484-8 Pa. $7.95

MYTHS OF THE HINDUS AND BUDDHISTS, Ananda K. Coomaraswamy and Sister Nivedita. Great stories of the epics; deeds of Krishna, Shiva, taken from puranas, Vedas, folk tales; etc. 32 illustrations. 400pp. 5⅜ x 8½. 21759-0 Pa. $12.95

THE TRAUMA OF BIRTH, Otto Rank. Rank's controversial thesis that anxiety neurosis is caused by profound psychological trauma which occurs at birth. 256pp. 5⅜ x 8½. 27974-X Pa. $7.95

A THEOLOGICO-POLITICAL TREATISE, Benedict Spinoza. Also contains unfinished Political Treatise. Great classic on religious liberty, theory of government on common consent. R. Elwes translation. Total of 421pp. 5⅜ x 8½. 20249-6 Pa. $9.95

MY BONDAGE AND MY FREEDOM, Frederick Douglass. Born a slave, Douglass became outspoken force in antislavery movement. The best of Douglass' autobiographies. Graphic description of slave life. 464pp. 5⅜ x 8½. 22457-0 Pa. $8.95

FOLLOWING THE EQUATOR: A Journey Around the World, Mark Twain. Fascinating humorous account of 1897 voyage to Hawaii, Australia, India, New Zealand, etc. Ironic, bemused reports on peoples, customs, climate, flora and fauna, politics, much more. 197 illustrations. 720pp. 5⅜ x 8½. 26113-1 Pa. $15.95

THE PEOPLE CALLED SHAKERS, Edward D. Andrews. Definitive study of Shakers: origins, beliefs, practices, dances, social organization, furniture and crafts, etc. 33 illustrations. 351pp. 5⅜ x 8½. 21081-2 Pa. $8.95

THE MYTHS OF GREECE AND ROME, H. A. Guerber. A classic of mythology, generously illustrated, long prized for its simple, graphic, accurate retelling of the principal myths of Greece and Rome, and for its commentary on their origins and significance. With 64 illustrations by Michelangelo, Raphael, Titian, Rubens, Canova, Bernini and others. 480pp. 5⅜ x 8½. 27584-1 Pa. $9.95

PSYCHOLOGY OF MUSIC, Carl E. Seashore. Classic work discusses music as a medium from psychological viewpoint. Clear treatment of physical acoustics, auditory apparatus, sound perception, development of musical skills, nature of musical feeling, host of other topics. 88 figures. 408pp. 5⅜ x 8½. 21851-1 Pa. $11.95

THE PHILOSOPHY OF HISTORY, Georg W. Hegel. Great classic of Western thought develops concept that history is not chance but rational process, the evolution of freedom. 457pp. 5⅜ x 8½. 20112-0 Pa. $9.95

THE BOOK OF TEA, Kakuzo Okakura. Minor classic of the Orient: entertaining, charming explanation, interpretation of traditional Japanese culture in terms of tea ceremony. 94pp. 5⅜ x 8½. 20070-1 Pa. $3.95

LIFE IN ANCIENT EGYPT, Adolf Erman. Fullest, most thorough, detailed older account with much not in more recent books, domestic life, religion, magic, medicine, commerce, much more. Many illustrations reproduce tomb paintings, carvings, hieroglyphs, etc. 597pp. 5⅜ x 8½. 22632-8 Pa. $12.95

SUNDIALS, Their Theory and Construction, Albert Waugh. Far and away the best, most thorough coverage of ideas, mathematics concerned, types, construction, adjusting anywhere. Simple, nontechnical treatment allows even children to build several of these dials. Over 100 illustrations. 230pp. 5⅜ x 8½. 22947-5 Pa. $8.95

THEORETICAL HYDRODYNAMICS, L. M. Milne-Thomson. Classic exposition of the mathematical theory of fluid motion, applicable to both hydrodynamics and aerodynamics. Over 600 exercises. 768pp. 6⅛ x 9¼. 68970-0 Pa. $20.95

SONGS OF EXPERIENCE: Facsimile Reproduction with 26 Plates in Full Color, William Blake. 26 full-color plates from a rare 1826 edition. Includes "TheTyger," "London," "Holy Thursday," and other poems. Printed text of poems. 48pp. 5¼ x 7. 24636-1 Pa. $4.95

OLD-TIME VIGNETTES IN FULL COLOR, Carol Belanger Grafton (ed.). Over 390 charming, often sentimental illustrations, selected from archives of Victorian graphics—pretty women posing, children playing, food, flowers, kittens and puppies, smiling cherubs, birds and butterflies, much more. All copyright-free. 48pp. 9¼ x 12¼. 27269-9 Pa. $7.95

PERSPECTIVE FOR ARTISTS, Rex Vicat Cole. Depth, perspective of sky and sea, shadows, much more, not usually covered. 391 diagrams, 81 reproductions of drawings and paintings. 279pp. 5⅜ x 8½. 22487-2 Pa. $7.95

DRAWING THE LIVING FIGURE, Joseph Sheppard. Innovative approach to artistic anatomy focuses on specifics of surface anatomy, rather than muscles and bones. Over 170 drawings of live models in front, back and side views, and in widely varying poses. Accompanying diagrams. 177 illustrations. Introduction. Index. 144pp. 8⅜ x11¼. 26723-7 Pa. $8.95

GOTHIC AND OLD ENGLISH ALPHABETS: 100 Complete Fonts, Dan X. Solo. Add power, elegance to posters, signs, other graphics with 100 stunning copyright-free alphabets: Blackstone, Dolbey, Germania, 97 more–including many lower-case, numerals, punctuation marks. 104pp. 8⅛ x 11. 24695-7 Pa. $8.95

HOW TO DO BEADWORK, Mary White. Fundamental book on craft from simple projects to five-bead chains and woven works. 106 illustrations. 142pp. 5⅜ x 8. 20697-1 Pa. $5.95

THE BOOK OF WOOD CARVING, Charles Marshall Sayers. Finest book for beginners discusses fundamentals and offers 34 designs. "Absolutely first rate . . . well thought out and well executed."–E. J. Tangerman. 118pp. 7¾ x 10⅝. 23654-4 Pa. $7.95

ILLUSTRATED CATALOG OF CIVIL WAR MILITARY GOODS: Union Army Weapons, Insignia, Uniform Accessories, and Other Equipment, Schuyler, Hartley, and Graham. Rare, profusely illustrated 1846 catalog includes Union Army uniform and dress regulations, arms and ammunition, coats, insignia, flags, swords, rifles, etc. 226 illustrations. 160pp. 9 x 12. 24939-5 Pa. $10.95

WOMEN'S FASHIONS OF THE EARLY 1900s: An Unabridged Republication of "New York Fashions, 1909," National Cloak & Suit Co. Rare catalog of mail-order fashions documents women's and children's clothing styles shortly after the turn of the century. Captions offer full descriptions, prices. Invaluable resource for fashion, costume historians. Approximately 725 illustrations. 128pp. 8⅜ x 11¼. 27276-1 Pa. $11.95

THE 1912 AND 1915 GUSTAV STICKLEY FURNITURE CATALOGS, Gustav Stickley. With over 200 detailed illustrations and descriptions, these two catalogs are essential reading and reference materials and identification guides for Stickley furniture. Captions cite materials, dimensions and prices. 112pp. 6½ x 9¼. 26676-1 Pa. $9.95

EARLY AMERICAN LOCOMOTIVES, John H. White, Jr. Finest locomotive engravings from early 19th century: historical (1804–74), main-line (after 1870), special, foreign, etc. 147 plates. 142pp. 11⅜ x 8¼. 22772-3 Pa. $10.95

THE TALL SHIPS OF TODAY IN PHOTOGRAPHS, Frank O. Braynard. Lavishly illustrated tribute to nearly 100 majestic contemporary sailing vessels: Amerigo Vespucci, Clearwater, Constitution, Eagle, Mayflower, Sea Cloud, Victory, many more. Authoritative captions provide statistics, background on each ship. 190 black-and-white photographs and illustrations. Introduction. 128pp. 8⅞ x 11¾. 27163-3 Pa. $14.95

LITTLE BOOK OF EARLY AMERICAN CRAFTS AND TRADES, Peter Stockham (ed.). 1807 children's book explains crafts and trades: baker, hatter, cooper, potter, and many others. 23 copperplate illustrations. 140pp. 4⅝ x 6.
23336-7 Pa. $4.95

VICTORIAN FASHIONS AND COSTUMES FROM HARPER'S BAZAR, 1867–1898, Stella Blum (ed.). Day costumes, evening wear, sports clothes, shoes, hats, other accessories in over 1,000 detailed engravings. 320pp. 9⅜ x 12¼.
22990-4 Pa. $15.95

GUSTAV STICKLEY, THE CRAFTSMAN, Mary Ann Smith. Superb study surveys broad scope of Stickley's achievement, especially in architecture. Design philosophy, rise and fall of the Craftsman empire, descriptions and floor plans for many Craftsman houses, more. 86 black-and-white halftones. 31 line illustrations. Introduction 208pp. 6½ x 9¼.
27210-9 Pa. $9.95

THE LONG ISLAND RAIL ROAD IN EARLY PHOTOGRAPHS, Ron Ziel. Over 220 rare photos, informative text document origin (1844) and development of rail service on Long Island. Vintage views of early trains, locomotives, stations, passengers, crews, much more. Captions. 8⅞ x 11¾.
26301-0 Pa. $13.95

VOYAGE OF THE LIBERDADE, Joshua Slocum. Great 19th-century mariner's thrilling, first-hand account of the wreck of his ship off South America, the 35-foot boat he built from the wreckage, and its remarkable voyage home. 128pp. 5⅜ x 8½.
40022-0 Pa. $4.95

TEN BOOKS ON ARCHITECTURE, Vitruvius. The most important book ever written on architecture. Early Roman aesthetics, technology, classical orders, site selection, all other aspects. Morgan translation. 331pp. 5⅜ x 8½. 20645-9 Pa. $8.95

THE HUMAN FIGURE IN MOTION, Eadweard Muybridge. More than 4,500 stopped-action photos, in action series, showing undraped men, women, children jumping, lying down, throwing, sitting, wrestling, carrying, etc. 390pp. 7⅞ x 10⅝.
20204-6 Clothbd. $27.95

TREES OF THE EASTERN AND CENTRAL UNITED STATES AND CANADA, William M. Harlow. Best one-volume guide to 140 trees. Full descriptions, woodlore, range, etc. Over 600 illustrations. Handy size. 288pp. 4½ x 6⅜.
20395-6 Pa. $6.95

SONGS OF WESTERN BIRDS, Dr. Donald J. Borror. Complete song and call repertoire of 60 western species, including flycatchers, juncoes, cactus wrens, many more–includes fully illustrated booklet. Cassette and manual 99913-0 $8.95

GROWING AND USING HERBS AND SPICES, Milo Miloradovich. Versatile handbook provides all the information needed for cultivation and use of all the herbs and spices available in North America. 4 illustrations. Index. Glossary. 236pp. 5⅜ x 8½.
25058-X Pa. $7.95

BIG BOOK OF MAZES AND LABYRINTHS, Walter Shepherd. 50 mazes and labyrinths in all–classical, solid, ripple, and more–in one great volume. Perfect inexpensive puzzler for clever youngsters. Full solutions. 112pp. 8⅛ x 11.
22951-3 Pa. $5.95

CATALOG OF DOVER BOOKS

PIANO TUNING, J. Cree Fischer. Clearest, best book for beginner, amateur. Simple repairs, raising dropped notes, tuning by easy method of flattened fifths. No previous skills needed. 4 illustrations. 201pp. 5⅜ x 8½. 23267-0 Pa. $6.95

HINTS TO SINGERS, Lillian Nordica. Selecting the right teacher, developing confidence, overcoming stage fright, and many other important skills receive thoughtful discussion in this indispensible guide, written by a world-famous diva of four decades' experience. 96pp. 5³/₈ x 8¹/₂. 40094-8 Pa. $4.95

THE COMPLETE NONSENSE OF EDWARD LEAR, Edward Lear. All nonsense limericks, zany alphabets, Owl and Pussycat, songs, nonsense botany, etc., illustrated by Lear. Total of 320pp. 5⅜ x 8½. (USO) 20167-8 Pa. $7.95

VICTORIAN PARLOUR POETRY: An Annotated Anthology, Michael R. Turner. 117 gems by Longfellow, Tennyson, Browning, many lesser-known poets. "The Village Blacksmith," "Curfew Must Not Ring Tonight," "Only a Baby Small," dozens more, often difficult to find elsewhere. Index of poets, titles, first lines. xxiii + 325pp. 5⅜ x 8¼. 27044-0 Pa. $8.95

DUBLINERS, James Joyce. Fifteen stories offer vivid, tightly focused observations of the lives of Dublin's poorer classes. At least one, "The Dead," is considered a masterpiece. Reprinted complete and unabridged from standard edition. 160pp. 5³/₁₆ x 8¼. 26870-5 Pa. $1.00

GREAT WEIRD TALES: 14 Stories by Lovecraft, Blackwood, Machen and Others, S. T. Joshi (ed.). 14 spellbinding tales, including "The Sin Eater," by Fiona McLeod, "The Eye Above the Mantel," by Frank Belknap Long, as well as renowned works by R. H. Barlow, Lord Dunsany, Arthur Machen, W. C. Morrow and eight other masters of the genre. 256pp. 5⅜ x 8½. (USO) 40436-6 Pa. $8.95

THE BOOK OF THE SACRED MAGIC OF ABRAMELIN THE MAGE, translated by S. MacGregor Mathers. Medieval manuscript of ceremonial magic. Basic document in Aleister Crowley, Golden Dawn groups. 268pp. 5⅜ x 8½. 23211-5 Pa. $9.95

NEW RUSSIAN-ENGLISH AND ENGLISH-RUSSIAN DICTIONARY, M. A. O'Brien. This is a remarkably handy Russian dictionary, containing a surprising amount of information, including over 70,000 entries. 366pp. 4½ x 6¼. 20208-9 Pa. $10.95

HISTORIC HOMES OF THE AMERICAN PRESIDENTS, Second, Revised Edition, Irvin Haas. A traveler's guide to American Presidential homes, most open to the public, depicting and describing homes occupied by every American President from George Washington to George Bush. With visiting hours, admission charges, travel routes. 175 photographs. Index. 160pp. 8¼ x 11. 26751-2 Pa. $11.95

NEW YORK IN THE FORTIES, Andreas Feininger. 162 brilliant photographs by the well-known photographer, formerly with *Life* magazine. Commuters, shoppers, Times Square at night, much else from city at its peak. Captions by John von Hartz. 181pp. 9¼ x 10¾. 23585-8 Pa. $13.95

INDIAN SIGN LANGUAGE, William Tomkins. Over 525 signs developed by Sioux and other tribes. Written instructions and diagrams. Also 290 pictographs. 111pp. 6⅛ x 9¼. 22029-X Pa. $3.95

CATALOG OF DOVER BOOKS

ANATOMY: A Complete Guide for Artists, Joseph Sheppard. A master of figure drawing shows artists how to render human anatomy convincingly. Over 460 illustrations. 224pp. 8⅜ x 11¼. 27279-6 Pa. $11.95

MEDIEVAL CALLIGRAPHY: Its History and Technique, Marc Drogin. Spirited history, comprehensive instruction manual covers 13 styles (ca. 4th century thru 15th). Excellent photographs; directions for duplicating medieval techniques with modern tools. 224pp. 8⅜ x 11¼. 26142-5 Pa. $12.95

DRIED FLOWERS: How to Prepare Them, Sarah Whitlock and Martha Rankin. Complete instructions on how to use silica gel, meal and borax, perlite aggregate, sand and borax, glycerine and water to create attractive permanent flower arrangements. 12 illustrations. 32pp. 5⅜ x 8½. 21802-3 Pa. $1.00

EASY-TO-MAKE BIRD FEEDERS FOR WOODWORKERS, Scott D. Campbell. Detailed, simple-to-use guide for designing, constructing, caring for and using feeders. Text, illustrations for 12 classic and contemporary designs. 96pp. 5⅜ x 8½. 25847-5 Pa. $3.95

SCOTTISH WONDER TALES FROM MYTH AND LEGEND, Donald A. Mackenzie. 16 lively tales tell of giants rumbling down mountainsides, of a magic wand that turns stone pillars into warriors, of gods and goddesses, evil hags, powerful forces and more. 240pp. 5⅜ x 8½. 29677-6 Pa. $6.95

THE HISTORY OF UNDERCLOTHES, C. Willett Cunnington and Phyllis Cunnington. Fascinating, well-documented survey covering six centuries of English undergarments, enhanced with over 100 illustrations: 12th-century laced-up bodice, footed long drawers (1795), 19th-century bustles, 19th-century corsets for men, Victorian "bust improvers," much more. 272pp. 5⅝ x 8¼. 27124-2 Pa. $9.95

ARTS AND CRAFTS FURNITURE: The Complete Brooks Catalog of 1912, Brooks Manufacturing Co. Photos and detailed descriptions of more than 150 now very collectible furniture designs from the Arts and Crafts movement depict davenports, settees, buffets, desks, tables, chairs, bedsteads, dressers and more, all built of solid, quarter-sawed oak. Invaluable for students and enthusiasts of antiques, Americana and the decorative arts. 80pp. 6½ x 9¼. 27471-3 Pa. $8.95

WILBUR AND ORVILLE: A Biography of the Wright Brothers, Fred Howard. Definitive, crisply written study tells the full story of the brothers' lives and work. A vividly written biography, unparalleled in scope and color, that also captures the spirit of an extraordinary era. 560pp. 6⅛ x 9¼. 40297-5 Pa. $17.95

THE ARTS OF THE SAILOR: Knotting, Splicing and Ropework, Hervey Garrett Smith. Indispensable shipboard reference covers tools, basic knots and useful hitches; handsewing and canvas work, more. Over 100 illustrations. Delightful reading for sea lovers. 256pp. 5⅜ x 8½. 26440-8 Pa. $8.95

FRANK LLOYD WRIGHT'S FALLINGWATER: The House and Its History, Second, Revised Edition, Donald Hoffmann. A total revision–both in text and illustrations–of the standard document on Fallingwater, the boldest, most personal architectural statement of Wright's mature years, updated with valuable new material from the recently opened Frank Lloyd Wright Archives. "Fascinating"–*The New York Times*. 116 illustrations. 128pp. 9¼ x 10¾. 27430-6 Pa. $12.95

PHOTOGRAPHIC SKETCHBOOK OF THE CIVIL WAR, Alexander Gardner. 100 photos taken on field during the Civil War. Famous shots of Manassas Harper's Ferry, Lincoln, Richmond, slave pens, etc. 244pp. 10⅝ x 8¼. 22731-6 Pa. $10.95

FIVE ACRES AND INDEPENDENCE, Maurice G. Kains. Great back-to-the-land classic explains basics of self-sufficient farming. The one book to get. 95 illustrations. 397pp. 5⅜ x 8½. 20974-1 Pa. $7.95

SONGS OF EASTERN BIRDS, Dr. Donald J. Borror. Songs and calls of 60 species most common to eastern U.S.: warblers, woodpeckers, flycatchers, thrushes, larks, many more in high-quality recording. Cassette and manual 99912-2 $9.95

A MODERN HERBAL, Margaret Grieve. Much the fullest, most exact, most useful compilation of herbal material. Gigantic alphabetical encyclopedia, from aconite to zedoary, gives botanical information, medical properties, folklore, economic uses, much else. Indispensable to serious reader. 161 illustrations. 888pp. 6½ x 9¼. 2-vol. set. (USO) Vol. I: 22798-7 Pa. $9.95
Vol. II: 22799-5 Pa. $9.95

HIDDEN TREASURE MAZE BOOK, Dave Phillips. Solve 34 challenging mazes accompanied by heroic tales of adventure. Evil dragons, people-eating plants, blood-thirsty giants, many more dangerous adversaries lurk at every twist and turn. 34 mazes, stories, solutions. 48pp. 8¼ x 11. 24566-7 Pa. $2.95

LETTERS OF W. A. MOZART, Wolfgang A. Mozart. Remarkable letters show bawdy wit, humor, imagination, musical insights, contemporary musical world; includes some letters from Leopold Mozart. 276pp. 5⅜ x 8½. 22859-2 Pa. $7.95

BASIC PRINCIPLES OF CLASSICAL BALLET, Agrippina Vaganova. Great Russian theoretician, teacher explains methods for teaching classical ballet. 118 illustrations. 175pp. 5⅜ x 8½. 22036-2 Pa. $5.95

THE JUMPING FROG, Mark Twain. Revenge edition. The original story of The Celebrated Jumping Frog of Calaveras County, a hapless French translation, and Twain's hilarious "retranslation" from the French. 12 illustrations. 66pp. 5⅜ x 8½.
22686-7 Pa. $3.95

BEST REMEMBERED POEMS, Martin Gardner (ed.). The 126 poems in this superb collection of 19th- and 20th-century British and American verse range from Shelley's "To a Skylark" to the impassioned "Renascence" of Edna St. Vincent Millay and to Edward Lear's whimsical "The Owl and the Pussycat." 224pp. 5⅜ x 8½.
27165-X Pa. $5.95

COMPLETE SONNETS, William Shakespeare. Over 150 exquisite poems deal with love, friendship, the tyranny of time, beauty's evanescence, death and other themes in language of remarkable power, precision and beauty. Glossary of archaic terms. 80pp. 5$\frac{3}{16}$ x 8¼. 26686-9 Pa. $1.00

BODIES IN A BOOKSHOP, R. T. Campbell. Challenging mystery of blackmail and murder with ingenious plot and superbly drawn characters. In the best tradition of British suspense fiction. 192pp. 5⅜ x 8½. 24720-1 Pa. $6.95

THE WIT AND HUMOR OF OSCAR WILDE, Alvin Redman (ed.). More than 1,000 ripostes, paradoxes, wisecracks: Work is the curse of the drinking classes; I can resist everything except temptation; etc. 258pp. 5⅜ x 8½. 20602-5 Pa. $6.95

SHAKESPEARE LEXICON AND QUOTATION DICTIONARY, Alexander Schmidt. Full definitions, locations, shades of meaning in every word in plays and poems. More than 50,000 exact quotations. 1,485pp. 6½ x 9¼. 2-vol. set.

Vol. 1: 22726-X Pa. $17.95
Vol. 2: 22727-8 Pa. $17.95

SELECTED POEMS, Emily Dickinson. Over 100 best-known, best-loved poems by one of America's foremost poets, reprinted from authoritative early editions. No comparable edition at this price. Index of first lines. 64pp. 5³⁄₁₆ x 8¼. 26466-1 Pa. $1.00

THE INSIDIOUS DR. FU-MANCHU, Sax Rohmer. The first of the popular mystery series introduces a pair of English detectives to their archnemesis, the diabolical Dr. Fu-Manchu. Flavorful atmosphere, fast-paced action, and colorful characters enliven this classic of the genre. 208pp. 5³⁄₁₆ x 8¼. 29898-1 Pa. $2.00

THE MALLEUS MALEFICARUM OF KRAMER AND SPRENGER, translated by Montague Summers. Full text of most important witchhunter's "bible," used by both Catholics and Protestants. 278pp. 6⅝ x 10. 22802-9 Pa. $12.95

SPANISH STORIES/CUENTOS ESPAÑOLES: A Dual-Language Book, Angel Flores (ed.). Unique format offers 13 great stories in Spanish by Cervantes, Borges, others. Faithful English translations on facing pages. 352pp. 5⅜ x 8½. 25399-6 Pa. $8.95

GARDEN CITY, LONG ISLAND, IN EARLY PHOTOGRAPHS, 1869–1919, Mildred H. Smith. Handsome treasury of 118 vintage pictures, accompanied by carefully researched captions, document the Garden City Hotel fire (1899), the Vanderbilt Cup Race (1908), the first airmail flight departing from the Nassau Boulevard Aerodrome (1911), and much more. 96pp. 8⅞ x 11¾. 40669-5 Pa. $12.95

OLD QUEENS, N.Y., IN EARLY PHOTOGRAPHS, Vincent F. Seyfried and William Asadorian. Over 160 rare photographs of Maspeth, Jamaica, Jackson Heights, and other areas. Vintage views of DeWitt Clinton mansion, 1939 World's Fair and more. Captions. 192pp. 8⅞ x 11. 26358-4 Pa. $12.95

CAPTURED BY THE INDIANS: 15 Firsthand Accounts, 1750-1870, Frederick Drimmer. Astounding true historical accounts of grisly torture, bloody conflicts, relentless pursuits, miraculous escapes and more, by people who lived to tell the tale. 384pp. 5⅜ x 8½. 24901-8 Pa. $8.95

THE WORLD'S GREAT SPEECHES (Fourth Enlarged Edition), Lewis Copeland, Lawrence W. Lamm, and Stephen J. McKenna. Nearly 300 speeches provide public speakers with a wealth of updated quotes and inspiration—from Pericles' funeral oration and William Jennings Bryan's "Cross of Gold Speech" to Malcolm X's powerful words on the Black Revolution and Earl of Spenser's tribute to his sister, Diana, Princess of Wales. 944pp. 5⅜ x 8⅜. 40903-1 Pa. $15.95

THE BOOK OF THE SWORD, Sir Richard F. Burton. Great Victorian scholar/adventurer's eloquent, erudite history of the "queen of weapons"—from prehistory to early Roman Empire. Evolution and development of early swords, variations (sabre, broadsword, cutlass, scimitar, etc.), much more. 336pp. 6⅛ x 9¼. 25434-8 Pa. $9.95

AUTOBIOGRAPHY: The Story of My Experiments with Truth, Mohandas K. Gandhi. Boyhood, legal studies, purification, the growth of the Satyagraha (nonviolent protest) movement. Critical, inspiring work of the man responsible for the freedom of India. 480pp. 5⅜ x 8½. (USO)   24593-4 Pa. $8.95

CELTIC MYTHS AND LEGENDS, T. W. Rolleston. Masterful retelling of Irish and Welsh stories and tales. Cuchulain, King Arthur, Deirdre, the Grail, many more. First paperback edition. 58 full-page illustrations. 512pp. 5⅜ x 8½.   26507-2 Pa. $9.95

THE PRINCIPLES OF PSYCHOLOGY, William James. Famous long course complete, unabridged. Stream of thought, time perception, memory, experimental methods; great work decades ahead of its time. 94 figures. 1,391pp. 5⅜ x 8½. 2-vol. set.
Vol. I: 20381-6 Pa. $13.95
Vol. II: 20382-4 Pa. $14.95

THE WORLD AS WILL AND REPRESENTATION, Arthur Schopenhauer. Definitive English translation of Schopenhauer's life work, correcting more than 1,000 errors, omissions in earlier translations. Translated by E. F. J. Payne. Total of 1,269pp. 5⅜ x 8½. 2-vol. set.
Vol. 1: 21761-2 Pa. $12.95
Vol. 2: 21762-0 Pa. $12.95

MAGIC AND MYSTERY IN TIBET, Madame Alexandra David-Neel. Experiences among lamas, magicians, sages, sorcerers, Bonpa wizards. A true psychic discovery. 32 illustrations. 321pp. 5⅜ x 8½. (USO)   22682-4 Pa. $9.95

THE EGYPTIAN BOOK OF THE DEAD, E. A. Wallis Budge. Complete reproduction of Ani's papyrus, finest ever found. Full hieroglyphic text, interlinear transliteration, word-for-word translation, smooth translation. 533pp. 6½ x 9¼.
21866-X Pa. $11.95

MATHEMATICS FOR THE NONMATHEMATICIAN, Morris Kline. Detailed, college-level treatment of mathematics in cultural and historical context, with numerous exercises. Recommended Reading Lists. Tables. Numerous figures. 641pp. 5⅜ x 8½.
24823-2 Pa. $11.95

PROBABILISTIC METHODS IN THE THEORY OF STRUCTURES, Isaac Elishakoff. Well-written introduction covers the elements of the theory of probability from two or more random variables, the reliability of such multivariable structures, the theory of random function, Monte Carlo methods of treating problems incapable of exact solution, and more. Examples. 502pp. 5³⁄₈ x 8¹⁄₂.   40691-1 Pa. $16.95

THE RIME OF THE ANCIENT MARINER, Gustave Doré, S. T. Coleridge. Doré's finest work; 34 plates capture moods, subtleties of poem. Flawless full-size reproductions printed on facing pages with authoritative text of poem. "Beautiful. Simply beautiful."—*Publisher's Weekly.* 77pp. 9¼ x 12.   22305-1 Pa. $7.95

NORTH AMERICAN INDIAN DESIGNS FOR ARTISTS AND CRAFTSPEOPLE, Eva Wilson. Over 360 authentic copyright-free designs adapted from Navajo blankets, Hopi pottery, Sioux buffalo hides, more. Geometrics, symbolic figures, plant and animal motifs, etc. 128pp. 8⅜ x 11. (EUK)   25341-4 Pa. $8.95

SCULPTURE: Principles and Practice, Louis Slobodkin. Step-by-step approach to clay, plaster, metals, stone; classical and modern. 253 drawings, photos. 255pp. 8⅛ x 11.
22960-2 Pa. $11.95

THE INFLUENCE OF SEA POWER UPON HISTORY, 1660–1783, A. T. Mahan. Influential classic of naval history and tactics still used as text in war colleges. First paperback edition. 4 maps. 24 battle plans. 640pp. 5⅜ x 8½. 25509-3 Pa. $14.95

THE STORY OF THE TITANIC AS TOLD BY ITS SURVIVORS, Jack Winocour (ed.). What it was really like. Panic, despair, shocking inefficiency, and a little heroism. More thrilling than any fictional account. 26 illustrations. 320pp. 5⅜ x 8½. 20610-6 Pa. $8.95

FAIRY AND FOLK TALES OF THE IRISH PEASANTRY, William Butler Yeats (ed.). Treasury of 64 tales from the twilight world of Celtic myth and legend: "The Soul Cages," "The Kildare Pooka," "King O'Toole and his Goose," many more. Introduction and Notes by W. B. Yeats. 352pp. 5⅜ x 8½. 26941-8 Pa. $8.95

BUDDHIST MAHAYANA TEXTS, E. B. Cowell and Others (eds.). Superb, accurate translations of basic documents in Mahayana Buddhism, highly important in history of religions. The Buddha-karita of Asvaghosha, Larger Sukhavativyuha, more. 448pp. 5⅜ x 8½. 25552-2 Pa. $12.95

ONE TWO THREE . . . INFINITY: Facts and Speculations of Science, George Gamow. Great physicist's fascinating, readable overview of contemporary science: number theory, relativity, fourth dimension, entropy, genes, atomic structure, much more. 128 illustrations. Index. 352pp. 5⅜ x 8½. 25664-2 Pa. $8.95

EXPERIMENTATION AND MEASUREMENT, W. J. Youden. Introductory manual explains laws of measurement in simple terms and offers tips for achieving accuracy and minimizing errors. Mathematics of measurement, use of instruments, experimenting with machines. 1994 edition. Foreword. Preface. Introduction. Epilogue. Selected Readings. Glossary. Index. Tables and figures. 128pp. 5³/₈ x 8¹/₂. 40451-X Pa. $6.95

DALÍ ON MODERN ART: The Cuckolds of Antiquated Modern Art, Salvador Dalí. Influential painter skewers modern art and its practitioners. Outrageous evaluations of Picasso, Cézanne, Turner, more. 15 renderings of paintings discussed. 44 calligraphic decorations by Dalí. 96pp. 5⅜ x 8½. (USO) 29220-7 Pa. $5.95

ANTIQUE PLAYING CARDS: A Pictorial History, Henry René D'Allemagne. Over 900 elaborate, decorative images from rare playing cards (14th–20th centuries): Bacchus, death, dancing dogs, hunting scenes, royal coats of arms, players cheating, much more. 96pp. 9¼ x 12¼. 29265-7 Pa. $12.95

MAKING FURNITURE MASTERPIECES: 30 Projects with Measured Drawings, Franklin H. Gottshall. Step-by-step instructions, illustrations for constructing handsome, useful pieces, among them a Sheraton desk, Chippendale chair, Spanish desk, Queen Anne table and a William and Mary dressing mirror. 224pp. 8⅛ x 11¼. 29338-6 Pa. $13.95

THE FOSSIL BOOK: A Record of Prehistoric Life, Patricia V. Rich et al. Profusely illustrated definitive guide covers everything from single-celled organisms and dinosaurs to birds and mammals and the interplay between climate and man. Over 1,500 illustrations. 760pp. 7½ x 10⅛. 29371-8 Pa. $29.95

Prices subject to change without notice.

Available at your book dealer or write for free catalog to Dept. GI, Dover Publications, Inc., 31 East 2nd St., Mineola, N.Y. 11501. Dover publishes more than 500 books each year on science, elementary and advanced mathematics, biology, music, art, literary history, social sciences and other areas.